T0174868

Muscular and Skeletal Anomalies in Human Trisomy in an Evo-Devo Context

Description of a T18 cyclopic fetus and comparison between Edwards (T18), Patau (T13) and Down (T21) syndromes using 3-D imaging and anatomical illustrations

Muscular and Skeletal Anomalies in Human Trisomy in an Evo-Devo Context

Description of a T18 cyclopic fetus and comparison between Edwards (T18), Patau (T13) and Down (T21) syndromes using 3-D imaging and anatomical illustrations

- Christopher M. Smith
- Janine M. Ziermann
- Corinne Sandone
- Edward T. Bersu
- Julia L. Molnar
- Marjorie C. Gondré-Lewis
- M. Ashraf Aziz
- Rui Diogo

CRC Press
Taylor & Francis Group
Boca Raton London New York

CRC Press is an imprint of the
Taylor & Francis Group, an **informa** business

A SCIENCE PUBLISHERS BOOK

Illustrations in the book by Christopher M. Smith and Julia L. Molnar

Cover illustrations: Background image: Illustrated by James Didusch (Mall, 1917). Used with permission. Department of Art as Applied to Medicine, Johns Hopkins University School of Medicine. Foreground images: Illustrations by Christopher M. Smith, Department of Art as Applied to Medicine, Johns Hopkins University School of Medicine.

CRC Press
Taylor & Francis Group
6000 Broken Sound Parkway NW, Suite 300
Boca Raton, FL 33487-2742

First issued in paperback 2019

© 2015 by Taylor & Francis Group, LLC
CRC Press is an imprint of Taylor & Francis Group, an Informa business

No claim to original U.S. Government works

ISBN-13: 978-1-4987-1137-1 (hbk)
ISBN-13: 978-0-367-37779-3 (pbk)

This book contains information obtained from authentic and highly regarded sources. Reasonable efforts have been made to publish reliable data and information, but the author and publisher cannot assume responsibility for the validity of all materials or the consequences of their use. The authors and publishers have attempted to trace the copyright holders of all material reproduced in this publication and apologize to copyright holders if permission to publish in this form has not been obtained. If any copyright material has not been acknowledged please write and let us know so we may rectify in any future reprint.

Except as permitted under U.S. Copyright Law, no part of this book may be reprinted, reproduced, transmitted, or utilized in any form by any electronic, mechanical, or other means, now known or hereafter invented, including photocopying, microfilming, and recording, or in any information storage or retrieval system, without written permission from the publishers.

For permission to photocopy or use material electronically from this work, please access www.copyright.com (http://www.copyright.com/) or contact the Copyright Clearance Center, Inc. (CCC), 222 Rosewood Drive, Danvers, MA 01923, 978-750-8400. CCC is a not-for-profit organization that provides licenses and registration for a variety of users. For organizations that have been granted a photocopy license by the CCC, a separate system of payment has been arranged.

Trademark Notice: Product or corporate names may be trademarks or registered trademarks, and are used only for identification and explanation without intent to infringe.

Visit the Taylor & Francis Web site at
http://www.taylorandfrancis.com

and the CRC Press Web site at
http://www.crcpress.com

Preface

The seemingly endless forms of life that inhabit our world inspire in us a deep-felt wonder. Could the variety we see before us be directed through mechanisms that can be illuminated through their own errors? The study of abnormal development, such as human trisomy and cyclopia, allows us to explore the mysteries and mechanisms behind normal evolutionary and developmental processes and can provide insight into how morphology changes throughout evolution. By studying the abnormal, we can determine the "normal" morphological and developmental mechanisms in comparison with the associated genetic conditions, better understand the correlations between phenotype and genotype, and explore the applications and implications of these data for medicine and public health. This book is one of the first of a new scientific area named "Evolutionary Developmental Anthropology"; specifically, in this book in human birth defects, rather than mutants of non-human model organisms, are studied to explore both normal and abnormal developmental and evolutionary mechanisms, processes and patterns. Moreover, this multidisciplinary work combines scientific research lead by Rui Diogo, a multi-awarded investigator, with state-of-the-art anatomical and medical illustrations and 3D imaging, done by two of the more brilliant—and also multiawarded—young illustrators in the US, Christopher Smith and Julia Molnar. By focusing on both muscular and skeletal birth defects in humans with trisomy and cyclopia, including Down syndrome—one of the most studied human syndromes that, due to its high incidence and the fact that individuals with this syndrome often live until adulthood, is of special interest to the scientific and medical community—, this book is of interest to a wide audience, including medical researchers, physicians, surgeons, medical and dental students, pathologists, and pediatricians, among others, while also being of interest to developmental and evolutionary biologists, anatomists, functional morphologists, and zoologists.

Acknowledgements

We would like to thank Valerie DeLeon, Sarah Poynton, David Rini, Gary Lees, and the Department of Art as Applied to Medicine at Johns Hopkins University School of Medicine for supporting the Masters thesis project leading to this book. A special thanks to pathologists William Green and Marie Fidélia-Lambert of the Howard University College of Medicine for identifying the specimen for our use and gathering critical initial data, Stephen Lin from the molecular imaging laboratory at Howard University College of Medicine for his help with CT imaging data and to the Vesalius Trust for providing funding for the initial case study research. This research was also supported by a start-up package to Rui Diogo, from the Howard University College of Medicine.

We also want to thank M. Chardon, B. Wood, B. Richmond, M. Ashley-Ross, P. Ahlberg, G. Wagner, C. Boisvert, M. Linde-Medina, J. Hutchinson, V. Abdala, S. Walsh, and P. Johnston for discussions about appendicular muscles. In addition, we would like to thank the very helpful comments of Sharlene Santana and Michael Alfaro about the parts of the book concerning evolutionary reversions and Dollow's law, and of Randall Roper about the parts regarding mouse models for Down syndrome and apoptosis. A special thanks to R. Walsh and F. Slaby (Department of Anatomy, George Washington University), R. Bernstein and S. McFarlin (Department of Anthropology, George Washington University), N. Rybczynski (Canadian Museum of Nature), H. Mays (Cincinnati Museum of Natural History), F. Pastor (Department of Anatomy, University of Valladolid), A. Gorow, H. Fitch-Snyder and B. Rideout (San Diego Zoo) and J. Fritz and J. Murphy (Primate Foundation of Arizona) for kindly providing the non-primate and primate mammalian specimens dissected during this project.

We offer our special gratitude to the following pioneering investigators of the field of gross and developmental anomalies associated with human and aneuploid syndromes: Dr. John M. Opitz; Dr. James C. Pettersen; Ms. Beverly A. Barash; Dr. Leonard Freedman; Dr. Sharon Colacino, and Dr. Jose L. Ramirez-Castro; our work endeavors to build upon the sound foundation constructed by these investigators. We also acknowledge the work of Mr. Erik O. Felix (Research Librarian) of the Louis Stokes Health Sciences Library of Howard University, Washington, DC; he helped us to locate difficult to find articles; Mr. Stafford Battle helped in the preparation of the manuscript.

Thank you to all of our family and friends. The first author of this book wants to specifically thank Dwight and Vira Smith, Matthew Smith, Harold and Jean Donato, Dwight and Carol Smith, and Katharine Jones for their love and constant support, as well as the Classes of 2014 and 2015 from the Department of Art as Applied to Medicine at the Johns Hopkins University School of Medicine, specifically James Abraham, Jackie Meyer, Katelyn McDonald, Michael Silver, Veronica Falconieri, and Samantha Welker for all of their inspiration, advice and encouragement throughout the creation of this project.

Contents

CHAPTER 1

Topics and Purpose of this Book

"There is grandeur in this view of life, with its several powers, having been originally breathed into a few forms or into one; and that, whilst this planet has gone cycling on according to the fixed law of gravity, from so simple a beginning endless forms most beautiful and most wonderful have been, and are being, evolved."

—*Charles Darwin*

1.1 Introduction

The seemingly endless forms of life that inhabit our world inspire in us a deep-felt wonder. Could the variety we see before us be directed through mechanisms hidden from sight, but illuminated through their own errors? The study of abnormal development, such as human trisomy, allows us to explore the mysteries and mechanisms behind normal evolutionary and developmental processes and can provide insight into how morphology changes throughout evolution. By studying the abnormal, we can determine the "normal" morphologic mechanisms in comparison with the associated genetic condition, better understand the correlations between phenotype and genotype, and explore the applications and implications of these data for medicine and public health.

1.2 The ontology, phylogeny and clinical importance of muscle variation seen in the light of the myology of human aneuploid syndromes

In the immediate aftermath of the publication of Darwin's pivotal *On the Origins of Species* (1859) the notion that the anatomical variation within a population was a manifestation of imperfect replication of a perfect "bauplan" (ground pattern) or pristine celestial template of a species became defunct. The fact that individuals constituting a population (species) varied from each other in most of their phenotypic traits was seen as *the* raw material on which natural selection acted to generate adaptation to the environment. Adaptation resulted by the process of differential reproduction of sexually breeding individuals in a population.

Instead of discounting variable traits in order to expose the underlying rigid morphological perfection, morphologists began to expose and inventory "imperfect" traits (i.e., variations) and assess their fitness (evolutionary) potential. Furthermore, comparative morphologists started to use variable phenotypic traits to distinguish kinship amongst species, i.e., to study their lineage. After the publication of Darwin's *The Descent of Man and Selection in Relation to Sex* (1871) there followed a near-explosion of studies documenting human muscular variations, including the occurrence of similar variations in the closest living relatives of humans—the Old World monkeys, the lesser apes (hylobatids: gibbons and siamangs), and the great apes of Africa (chimpanzees and gorillas) and Asia (orangutans).

By the end of the 19th century, several annotated compilations containing documented muscle variations found in numerous dissected humans studied in leading European medical schools, museums and related research centers were published (e.g., Macalister, 1875; Ruge, 1887; Testut, 1888; Chudzinski, 1896; Le Double, 1897). In addition to these compendia, Macalister (1866; 1867) and Wood (1864; 1865; 1866; 1867a,b; 1868; 1869; 1870) wrote numerous papers regarding their findings of muscular variations in dissected humans, monkeys and apes.

Inspired by Darwin's works, comparative morphologists also began to compare visually recognizable human populations (so called "races") to ascertain their evolutionary relationships, i.e., they attempted to develop a hierarchical lineage of extant human populations. Disregarding contrary evidence (it was commonly known that various human "races" were actually interfertile), leading European anatomists developed (provisional) diagnostic profiles of musculature found in different human "races". The notion that the human species was clearly differentiated into scientifically definable "races" *which acted like distinct species* (i.e., they acted like reproductively isolated populations) began to emerge leading to the publication of the so-called "racial anatomical" studies.

One particular muscle group—the muscles of facial expression (mimetic muscles)—received close scrutiny to discern so-called "racial" differences and "hierarchy" in human groups (Chudzinski, 1896; Futamura, 1906; 1907; Loth, 1912; 1950; Lightoller, 1928; Huber, 1931; for further references see Huber, 1931). It is possible that these studies used Darwin's brilliant yet speculative text *The Expression of the Emotions in Man and Animals* (1872) as their inspiration. Huber (1931) summarized nearly a half century of descriptive and analytical work conducted by leading European anatomists on the mimetic muscles of human and nonhuman primates in order to place them in an "ascending scale" of primates and of humans of different "races".

However, recent reinvestigations of the comparative morphology of facial muscles using micro-dissection and other sophisticated, precise techniques (e.g., face mask removal and dissection) have shown that the earlier claims of clear-cut "racial" profiles and hierarchy were spurious at best, and even outright erroneous (Burrows et al., 2006; Burrows et al., 2009; Diogo et al., 2009b). Nevertheless, the subject of degrees of separation of different human "races" and their claimed hierarchy continues to receive attention in the popular imagination. Recently, a number of best selling books on the subject have appeared in the United States (Epstein, 2013; Wade, 2014; see also Gould, 1993).

Therefore, the proper and detailed documentation and study of muscular variations is a topic of great and urgent ontogenetic, phylogenetic, medical, and even social import. Before World War II most authoritative anatomy textbooks in English (e.g., Cunningham's, Gray's, Morris', Piersol's, Quain's) included lists of significant regional muscular variations in humans. The reason that this information was included in these textbooks was that muscle variability was considered to be an important clinical issue; surgeons were especially alerted to the existence of such variations in the human body. Following the war, textbooks began to phase out mention of such variations except in rare cases. Indeed, newer textbooks have now ceased to categorize muscles based on their ontogenic and phylogenic relationships; muscles are organized on the basis of their utilitarian (clinical) value. Contemporary books which are exclusively designed (in the United States of America) to provide the shortest road map to success in the National Board of Medical Examiners (NBME) Part I tests exclude mention of muscle (and other) variations altogether (Snell, 1995; Drake et al., 2014). The human form has been distilled into a non-variable data set (Aziz and McKenzie, 1999).

However, even a cursory inspection of leading contemporary anatomical research journals (e.g., *Journal of Anatomy; The Anatomical Record* and, especially *Clinical Anatomy*) show that muscle variations are of significant clinical importance [N.B., Wood (1868) estimated that each human body contains no fewer than ten muscle variations]. Such variations are clinically relevant for the following documented reasons: misdiagnosis of supernumerary muscles (which are confused with tumors); inexplicable entrapment neuropathies; compromised blood supply due to thrombosis; blood clot formation due to compression of major veins, etc. (Table 1). With the introduction of high-resolution medical diagnostic imaging techniques (e.g., CT and MRI) many of these muscular variations are often seen in these images; they confound the inadequately trained radiologists. Without a sound grounding in the knowledge of muscle variations of individual patients, the likelihood of misdiagnosis and or incorrect treatment (especially during surgical procedures) increases. Of course, many such misdiagnoses and/or surgical mistakes have serious legal consequences (Kohn et al., 2000). Furthermore, reconstructive microsurgery has progressed so far that amputated digits and other parts of the hand (even the forearm) can now be reattached and repurposed; success of these procedures demands a detailed knowledge of the anatomy of the hand, especially its variable musculature (Chung, 2012).

Comprehensive and detailed dissections of normal human embryos and fetuses by Cihak and his associates (at Charles University, Prague, Czech Republic; see Cihak, 1972) have revealed that, during normal human embryogenesis many muscles, including those which were regarded to be "atavistic" (evolutionary relics), appear during development only to be resorbed prior to parturition. During myogenesis, most individual muscles undergo partitioning to form muscle slips, many of which are resorbed prior to birth, leaving only the main "normal" muscles. Therefore, the sporadic occurrence of at least some supernumerary muscles in both karyotypically normal (e.g., variations) and abnormal (e.g., birth defects) human individuals is due to the fact that the extra slips where not resorbed during the fetal stage, which might in turn be due to developmental delay (see Chapter 6.3).

Importantly for this present work, delayed development of at least some organs—including muscles—is a diagnostic feature of all human aneuploid syndromes. The study of trisomic embryos, fetuses and neonates thus provides us with magnified clues to the complex history of individual muscles during development and also allows us to chart the development of variable human phenotypes more accurately.

1.3 Table 1—Examples of muscle variations and their clinical correlations in karyotypically normal humans

Date	Authors	Anomalous Muscle(s)	Clinical Manifestation(s)
1979	Stark et al.	Contrahentes digitorum	Carpal Tunnel Syndrome
1997	Harry et al.	Scalenus minimus	Compression of inferior trunk of brachial plexus; Thoracic Outlet Syndrome
1997	Fabrizio and Clemente	Triceps brachii with 4th head	Radial nerve palsy; vascular compromise due to compression of deep radial artery; snapping elbow

Table 1 contd....

Table 1 contd.

Date	Authors	Anomalous Muscle(s)	Clinical Manifestation(s)
1998	Wahba et al.	Accessory flexor digiti minimi profundus	Ulnar nerve palsy due to compression, Guyon's Canal; ulnar artery thrombosis
1998	Nakatani et al.	4-headed biceps brachii	Median nerve palsy; brachial artery compression
1998	Pessa et al.	Bifid zygomaticus major	Formation of cheek dimple
1998	Penhall et al.	Pterygoideus proprius	Maxillary artery compression
2000	Oh et al.	Flexor pollicus longus accessory head	Anterior interosseous nerve palsy; compromised precision grip
2000	Rosenheimer et al.	Levator claviculae	This muscle causes cervical lymphadenopathy
2001	Schön-Ybarra and Bauer	Medial portion of temporalis	Tic douloureux; maxillary nerve compression
2001	Forcada et al.	Subclavius posticus	Thoracic Outlet Syndrome
2001	El-Naggar and Zahir	Coracobrachialis with 3 bellies; 3rd head of biceps brachii	Musculocutaneous nerve compression
2002	Scott-Conner and Al-Jurf	Sternalis; pectoralis quartus; axillary arch muscle	Extra muscles in radical mastectomy
2002	Rodriguez-Niedenführ et al.	Extensor digiti brevis manus	Dorsal wrist ganglion; source of tendon transfer
2002	Ragoowansi et al.	Musculus transversus carpi	Carpal Tunnel Syndrome
2002	Bonastre et al.	Pectoralis quartus; axillary arch muscle	Axillary thrombosis; complicates removal of axillary lymph nodes
2003	Kobayashi et al.	Anomalous muscle associated with flexor digitorum superficialis	
2003	Madhavi and Holla	Anomalous flexor digiti minimi brevis	Compression of ulnar nerve and artery in Guyon's Canal; Ulnar Tunnel Syndrome
2003	Arraéz-Aybar et al.	Sternalis	Complication of breast surgery; alters ECG readings
2003	Merida-Velasco et al.	Axillary arch muscle	Compression of brachial plexus cords; hyperabduction syndrome; axillary thrombosis
2004	El-Naggar and Al-Saggaf	Coracobrachialis longus	Compression of median nerve, brachial artery and median nerve; palsy hypoxia of forearm and hand
2005	Turgut et al.	Axillopectoral (Langer's or Axillary arch) muscle	Compression of brachial plexus cords and branches; axillary thrombosis
2005	Plock et al.	Levator palpebrae superioris accessorius	Ptosis
2005	Tubbs et al.	Contrahentes digitorum	Carpal Tunnel Syndrome
2006	Loukas et al.	Accessory brachialis	Median nerve palsy
2006	Tiengo et al.	Accessory palmaris longus	Emulates tumor; median nerve compression; Carpal Tunnel Syndrome
2006a	Tubbs et al.	Tensor fasciae suralis	Appears as a mass in popliteal fossa

Table 1 contd....

Table 1 contd.

Date	Authors	Anomalous Muscle(s)	Clinical Manifestation(s)
2006c	Tubbs et al.	Triceps brachii 4th head	Radial nerve palsy; profunda brachii artery compression
2006b	Tubbs et al.	Psoas quartus m.	Femoral nerve compression; weak patellar tendon reflex
2007	Georgiev et al.	Axillary arch (Langer's) muscle	Axillary swelling; axillary thrombosis; brachial plexus compression
2008	Jay et al.	Plantaris	Patellofemoral pain syndrome
2008	Pai et al.	Flexor pollicis longus and flexor digitorum profundus (Gantzer's) accessory heads	Median nerve and/or anterior interosseous nerve palsy
2008	Ammendolia	Extensor digitorum brevis manus	Tumor; dorsal wrist ganglion

1.4 Trisomies 18, 13, and 21, cyclopia, and lack of comparative myological studies

Trisomy is often caused by a perturbation of meiosis II in the mother's eggs, in which a failure of sister chromatids to separate leaves the gamete with two instead of the normal one chromatid for a given chromosome. This extra chromosome can produce many different phenotypic outcomes throughout the body. In this book, we will investigate the muscular and skeletal abnormalities observed in a 28-week human Trisomy 18 cyclopic fetus and compare this individual with other humans with Trisomy 18 (Edwards syndrome) as well as with Trisomy 13 (Patau syndrome) and Trisomy 21 (Down syndrome). Our observations, comparisons, and review of the literature will allow us to examine and discuss possible similarities and differences from the individual to the syndrome level. Our research also has other significant associated objectives: to delineate the muscle anomalies caused by aneuploid syndromes; to identify those muscular traits which are diagnostic for each one of the common aneuploid syndromes; to help chart the morphogenetic pathways of these diagnostic anomalies; and, to assess the variability (descriptive and quantitative) shown by the diagnostic muscular defects in the different aneuploid syndromes.

Trisomy 18 is the second most prevalent autosomal trisomy after Trisomy 21. Humans with Trisomy 18 usually do not survive more than one year after birth (Bugge et al., 1998; Cereda and Carey, 2012). This syndrome causes a number of phenotypic alterations, including overlapping digits in the hand (Fig. 1.1) and modified cranial morphology. Trisomy 21, or Down syndrome, is the most common and well-known human trisomy. Individuals who have this syndrome, which is characterized by a flattened face and upward-slanting eyes, can live well into adulthood. Trisomy 13 is the least common of the three trisomies and often results in multiple anomalies including malformed ears and cleft palates. One consequence of chromosomal trisomy is the possibility of severe craniofacial malformations. One of the most severe is alobar holoprosencephaly, or cyclopia, which is a complete failure of the two cerebral hemispheres to separate in development (Fig. 1.2). As a result of the fused cerebral hemispheres, the eyes are either partially or completely fused. DeMyer et al. (1964) suggested the phrase "the face predicts the brain" to illustrate how the developing face follows the developing brain, meaning that if there is a failure of the two halves of the brain to separate, the face will reflect this pattern. This process typically occurs during the third week of gestation, before generation of eye anlagen and bilateral cleavage of the developing forebrain (Garzozi and Barkay, 1985). In many cyclopia cases, a proboscis is located above the median eye. According to McGrath (1992), "the proboscis in human cyclopia approximates to the anterosuperior part of the nasal cavity developed in the absence of median components." The presence of a proboscis is thought to be due to the failure of the eyes to separate, which prevents the nose structures from migrating down the face early in development (e.g., Leroi, 2005).

Trisomy 18 and cyclopia have been found in the same individual, albeit rarely, and this combination has not been thoroughly described. The musculature in cyclopia is poorly documented (but see Mieden, 1982), and there are no myological descriptions of cyclopic Trisomy 18 individuals. The lack of muscular studies reflects the unfortunate neglect in recent decades of soft tissues such as muscles in both "normal" and abnormal phenotypes (Diogo and Wood, 2012a). This book is part of a long-term project aimed at contributing to the renaissance of comparative anatomy in general and comparative myology in particular (including the work of Bernard Wood and others listed in the Acknowledgements, as well as many who could not be directly thanked or mentioned in this book). This project was also inspired by the rising need for comparative studies between vertebrate model organisms in the relatively new and increasingly important field of evolutionary developmental biology (evo-devo). We hope that the observations, comparisons, and literature review provided in this book will contribute to a better understanding of both "normal" and abnormal development, evolution, and birth defects.

1.5 Order *versus* randomness in evolution and birth defects

Researchers such as the late Pere Alberch have theorized about how constraints in development and evolution might limit the variety of morphologies possible for a given organism (Alberch, 1989). He argued that, for a given developmental perturbation, only a certain number of resulting phenotypes are possible. According to Alberch, there is a logic or inherent "order" even in severe birth defects and alterations of developmental processes, a concept which he unfortunately designated the "logic of monsters." According to this theory, phenotypic birth defects caused by completely different genetic syndromes would tend to be similar to each other and also to natural variations in the "normal" phenotypes present in other taxa. A contrasting theory was proposed by Burton Shapiro, who has written extensively on homeostasis and the balance of developmental programs and buffering involved in genetic interactions (Shapiro, 1983). According to Shapiro's theory, once a developmental perturbation arises, the organism is universally disrupted and interactions between genes result in *random* and *chaotic* phenotypic effects, creating more intense changes in less "buffered" traits.

In order to test the "logic" and "randomness" theories, one must be able to observe their effects in the natural world. The detailed musculoskeletal study of human trisomies can provide novel scientific knowledge in this area and empirically test these two ways of interpreting anatomical birth defects. These studies also contribute to a better understanding of the development and evolution of our species and have the potential to improve surgical treatment for craniofacial and limb abnormalities.

1.6 Serial homology, integration, forelimbs and hindlimbs

The study of birth defects, including human trisomy, can also shed light upon the integration of and serial homology between the forelimbs and hindlimbs. The idea that these limbs are evolutionarily integrated has been long examined (Rolian et al., 2010) and may apply whether or not one sees the two limbs as serial homologues, as almost all authors do. For instance, it has been proposed that in our evolutionary bipedal lineage hands and feet became less integrated than they are in quadrupedal primates, and that this increasing modularity has allowed for different functional morphologies despite both sets of limbs sharing at least some similar genetic and developmental blueprints (Rolian, 2009). However, these studies largely focused on skeletal structures of wildtype animals, and therefore the combination of data from both skeletal *and* muscular structures and from both wildtype non-human animals *and* human individuals with and without birth defects may shed further light on this and related questions. In fact, it was the inclusion of muscles in a recent extensive comparative study of the pelvic and pectoral appendages of fish and the fore- and hindlimbs of tetrapods that led Diogo et al. (2013a) to propose that the pectoral and pelvic appendages are not serial homologues, a hypothesis that goes against a two-century-old dogma of fore- and hindlimb serial homology. Accordingly, the pectoral and pelvic appendages would have been anatomically different in ancestral lobe-finned fishes and only later become more similar to each other when they were respectively transformed into the fore- and hindlimbs of tetrapods, due to derived co-option of similar genes used in the development of both limbs.

1.7 Developmental constraints, muscle attachments, facial muscles, and the present study

A related theory concerns developmental constraints and the spatial associations between muscles and bones. The topological position (adult spatial relationships) of muscles during limb development has been shown to play a role in their attachments. In a previous study, we investigated the changes in origins and insertions of muscles in a Trisomy 18 fetus compared to karyotypically normal individuals (Diogo et al., in press). In one of these cases, a six-digit hand of a Trisomy 18 neonate with two thumbs (preaxial polydactyly), the more ulnar hand muscles that normally attach onto the thumb (e.g., adductor pollicis) attached to the more ulnar thumb, and the more radial hand muscles that normally attach onto the thumb (e.g., abductor pollicis brevis) attached to the more radial thumb (Fig. 1.3). The other hand of the same individual was missing the first digit (thumb), but, remarkably, all of the thumb muscles were present, attaching instead to the index digit. In the absence of a thumb, the muscle insertions shifted to the next nearest digit, based on their relative positions in space (Fig. 1.4). This pattern of attachment can be described as "nearest neighbor" (muscles attach to the closest developing bones/cartilage) as opposed to "seek and find" [muscle attachment is determined by developmental anlage (embryonic precursor group of cells) or homeotic identity], because the index digit of the 4-digit hand develops from the anlage, and has the homeotic identity, of digit 2, not digit 1.

Here we provide a detailed description of the musculoskeletal system of a Trisomy 18 cyclopic fetus and comparison with other individuals with Trisomies 18, 13, and 21. In addition to the theories outlined in the paragraphs above, these data will allow us to partially test the hypothesis that, in the head, attachments of most muscles are constrained by the developmental anlage of the branchial arches and neural crest cell populations rather than topological position, as proposed by Köntges and Lumsden (1996). Specifically, we will investigate whether, in trisomic individuals with muscle defects, changes in the head musculature are more integrated with the changes in the skull, relying mainly on pre-destined attachment points between specific muscles and specific bones of the same arch. We will pay special attention to the muscles of facial expression because their developmental program seems to be more similar to that of limb muscles than to that of other branchiomeric muscles (i.e., branchial arch muscles). For instance, the muscles of facial expression are migratory muscles and the gene *Met* contributes to their development (*Met* does not contribute to the development of other branchiomeric muscles, but it contributes to the development of limb muscles and hypobranchial muscles such as tongue and infrahyoid muscles, which are derived from somites; Prunotto et al., 2004; see Fig. 1.5).

In addition to elucidating these and other broader developmental and evolutionary questions, investigation into human trisomies can provide an anatomical basis for surgical procedures on variant human anatomy. As explained above, human individuals with Trisomy 18 usually do not survive long after birth. However, these individuals may share similar phenotypic patterns with individuals with other syndromes, including those with Trisomy 21, who often survive well into adulthood. If such similar patterns are found, studies like this one may contribute to human medicine by furthering our understanding of birth defects and informing surgical procedures on babies and children with Down syndrome and/or non-pentadactyl limbs (see Section 1.5 first paragraph). We hope that the illustrations and 3D reconstructions included in this book will contribute significantly to the understanding of the complex spatial relationships and attachments of the musculature. These illustrations will also help surgeons involved in reconstructive surgery to gain significant morphological insights to conduct successful repairs of defects. For this purpose, we used programs such as *Osirix*, *123DCatch* and *Zbrush* to aid in reconstructing musculature and thus more clearly describing the anatomy of individuals with severe birth defects. For further information on the techniques used to reconstruct these dissections, please contact the first author, Christopher M. Smith. The Trisomy 18 cyclopic human fetus reported in detail in Chapter 2 was provided to Marjorie Gondre-Lewis by pathologists William Green and Marie Fidélia-Lambert of the Howard University College of Medicine, and we estimate its age to be 28 weeks post coitum; for further information please contact Marjorie Gondre-Lewis.

The Musculoskeletal System of a 28-week Human Trisomy 18 Cyclopia Fetus

2.1 Introduction

Anatomical observations are presented in the order in which the dissection was undertaken, with particular attention to differences between the muscular phenotype of the fetus and that of karyotypically normal human fetuses (attachments and overall configuration of muscles are phenotypically normal unless explicitly noted otherwise). All muscle anomalies observed in this fetus are listed in Table 2 and compared with any previous reports found in literature on individuals with Trisomies 18, 13, or 21. The fetus' internal organs and the infrahyoid, thoracic, abdominal, and anterior pelvic structures, including the muscles, were removed prior to our study. Dissection began with removal of the skin on the back and shoulders using forceps, scalpel, and scissors. Fascia covering the musculature was carefully removed and muscle bellies were cleaned and identified. The dissection proceeded in the following order: upper limbs, trunk, pelvis, lower limbs, hands, feet, head, and neck. The following traits of each muscle were noted: presence/ absence, anatomical origin(s) and insertion(s), and number of bellies and/or tendons. The information provided in the text of this Chapter is complemented by Table 2 and Figs. 2.1–2.17 and 4.1, as well as the figures in Appendices A and B.

2.2 Back, shoulder and arm (Fig. 2.1)

We began the dissection by removing the skin of the back and posterior shoulders. The superficial muscles were observed, and no obvious anomalies were noted. Upon reflection of the **trapezius**, the **rhomboideus major** and **rhomboideus minor**, which were fused bilaterally (i.e., on both sides of the body), appeared to be comprised of 1/3 muscle fibers and 2/3 tendon/aponeurosis, while their origins from the spinous processes extended slightly higher than usual. This condition contrasts with that usually found in karyotypically normal humans, in whom both rhomboideus muscles are mostly composed of muscle fibers. In the neck, trapezius and **splenius capitis** were tightly associated at the midline origin, bilaterally.

The shoulder girdle showed minimal abnormal morphology. The only confirmed anomaly was the fusion of **latissimus dorsi** and **teres major** toward their humeral insertion on the medial lip of the intertubercular sulcus. On the left side of the body, **pectoralis major** was completely fused to **deltoideus,** creating a "deltopectoral complex" (Aziz, 1979; 1981a,b). On the right side, pectoralis major was only partially fused with deltoideus towards its insertion onto the humerus. A peculiar tendon arose from extra fibers on the anterior-most portion of biceps and inserted into the abdominal fibers of pectoralis major, linking the two muscles. The bicipital aponeurosis of biceps brachii was absent bilaterally, and the short head of biceps was missing on the left side. In addition, an anomalous deep head of **coracobrachialis** (profundus) was present on the left side. Coracobrachialis on the right side was fused to the short head of biceps and inserted through its entire length along the anteromedial proximal humerus instead of normally inserting at its termination.

2.3 Left forearm/hand (Figs. 2.2–2.4, 4.1)

Dissection proceeded down the left arm and into the hand. A dissection microscope was used to allow finer dissection and easier differentiation between tissues. **Pronator teres** and **flexor carpi radialis** were fused, while a second deep belly of pronator teres was present with a normal origin and a broader insertion that extended more distally onto the radius. **Extensor carpi ulnaris, extensor digit minimi**, and **extensor digitorum** were all fused proximally. Most other forearm extensors had slight fusions towards their origins, including **extensor carpi radialis longus**, **extensor carpi radialis brevis** and **brachioradialis. Palmaris longus** and **palmaris brevis** were both absent. **Flexor digitorum superficialis** originated only from the common flexor tendon and proximal ulna, as it was missing its normal radial origin. The tendon of flexor digitorum superficialis to digit 5 was absent, while the tendon of **flexor digitorum profundus** inserted normally onto the distal phalanx of digit 5. This configuration was mirrored on the dorsal aspect of the limb; the extensor digitorum tendon to digit 5 also was absent, and extensor digiti minimi was the sole extensor of digit 5. Extensor digiti minimi sent an intermediate tendon to digit 4 as well, but no tendon could be identified coming from extensor digitorum of digit 4 to digit 5. The portion of flexor digitorum profundus that was devoted to digit 5, however, was composed almost entirely of tendon. Upon reflection of the superficial flexors, a muscle slip was observed arising from flexor digitorum superficialis and part of **flexor carpi radialis** to insert onto **flexor pollicis longus,** the two tendons merging together. **Extensor pollicis brevis** was diminutive and fused with **abductor pollicis longus.**

In the left hand, the tendon of abductor pollicis longus trifurcated, sending a tendon to its usual insertion point on metacarpal 1 and two tendons to an unusual thenar mass of muscles. This mass was composed of what would normally be **opponens pollicis, flexor pollicis brevis** and **abductor pollicis brevis.** Upon further dissection of the thenar area, a peculiar muscle belly, which we will refer to as "**musculus interosseus accessorius,**" was found originating from the radial base of the proximal phalanx of digit 2 and inserting onto the radial base of the proximal phalanx of digit 1, partially fusing with the insertion of **adductor pollicis.** The oblique head of adductor pollicis was separated into distinct fiber bundles instead of forming a cohesive belly. When the thenar muscle mass was removed, an **adductor pollicis accessorius** (also referred to as "**interosseous volaris primus of Henle**") muscle was present, arising from the proximal portion of metacarpal 1 and inserting onto the base of proximal phalanx 1. Its insertion blended with that of adductor pollicis. Although adductor pollicis accessorius is usually not shown in human anatomical atlases, this muscle is actually found in most modern humans (Diogo and Wood, 2011). The 1st of the **lumbricales** (which usually attaches to digit 2) was absent, and flexor digitorum profundus to digit 5 did not contribute to the lumbricalis attaching to the 5th digit.

2.4 Right forearm/hand (Figs. 2.5–2.7, 4.1)

In the right arm, pronator teres had the same anomalous configuration as in the left arm, with an additional deep head inserting more distally on the radius. The more superficial pronator teres belly was fused along its entire length with flexor carpi radialis. Extensor carpi radialis longus and brevis were fused proximally. The **extensor indicis** tendon bifurcated at the metacarpal-phalangeal joint and inserted onto digits 2 and 3, instead of digit 2 only as in karyotypically normal humans (see Aziz and Dunlap, 1986). As in the left forearm, the right forearm had a muscular connection between the flexors. However, on the right side this connection was doubled, originating from both flexor carpi radialis and the superficial head of pronator teres and inserting onto flexor pollicis longus, blending with its tendon. As in the left forearm, flexor digitorum superficialis originated only from the ulna and the common flexor tendon. A separate belly of flexor digitorum superficialis sent a tendon to digit 2 and fused only proximally with the rest of the muscle.

In the right hand, extensor digitorum and flexor digitorum superficialis were strikingly similar: both were missing the tendon to digit 5, as in the left hand, and the intermediate tendon of extensor digitorum spanned digits 3 to 4 and 4 to 5 only. In the right hand, **extensor pollicis brevis** was completely missing, whereas it was present but diminutive in the left hand. The **extensor pollicis longus** tendon had the same structure as in the left hand, where it split in two at the level of the metacarpal and fused again before its normal insertion. In the thenar area, opponens pollicis, abductor pollicis brevis, and flexor pollicis brevis were completely absent. Instead, flexor pollicis longus had a tendon which bifurcated around the base of the first metacarpal to insert on the bases of both the proximal and distal phalanges of digit 1. The peculiar "musculus interosseus accessorius" found in the thenar area of the left hand was also observed in the right hand. It originated from the radial base of metacarpal 2 and inserted onto the radial base of metacarpal 1. Adductor pollicis accessorius ("interosseous volaris primus of Henle") muscle was also present in this hand, arising from the proximal portion of metacarpal 1 and inserting into the base of proximal phalanx 1. Its insertion blended with that of adductor pollicis, as in the left hand. Also as in the left hand, the 1st lumbricalis was absent, and flexor digitorum profundus to digit 5 did not contribute to any lumbricalis.

2.5 Legs and feet (Fig. 4.1)

Moving down the body, we noted some anomalies in the legs and feet but not in the pelvis or thighs. On the right side of the body, **popliteus** was missing, **gastrocnemius** and **soleus** were fused toward their insertion onto the Achilles tendon, and **fibularis tertius** had a separate belly rather than fusing, as normal, with **extensor digitorum longus**. The separate belly gave off a tendon to extensor digitorum longus proximally; the tendon of fibularis tertius and extensor digitorum longus tendon to the 4th digit also exchanged tendons. Also on the right side of the body, there was no extensor digitorum longus tendon to digit 5, but the tendinous contribution from extensor digitorum longus to fibularis tertius extended onto the distal phalanx of the 5th digit. The **extensor digitorum brevis** tendon to digit 2 bifurcated and sent tendons to digits 2 and 3. On the plantar side of the foot, the **flexor digitorum brevis** tendon to digit 5 was absent and the tendon to digit 4 was diminutive. In addition, the transverse head of **adductor hallucis** was absent.

In the left leg and foot, the gastrocnemius and soleus bellies were fused for the majority of their length. The flexor digitorum brevis tendon to digit 5 was absent, and the tendon to digit 4 was diminutive, similar to the right foot.

2.6 Neck and head, including extraocular muscles (Figs. 2.8–2.14)

Because most of the anterior neck was removed during a previous dissection, much of the infrahyoid musculature was unobservable. However, some of the suprahyoid muscles, mainly on the left side, were intact enough for their attachments to be observed. The **sternocleidomastoideus** attachment on the skull was very broad, and the belly itself consisted of many separate fascicles. This muscle's attachment also was more continuous than normal with **trapezius**, which itself had a very broad insertion onto the skull.

Regarding the muscles of facial expression, a strong fascial connection between **platysma myoides** and the neck muscle sternocleidomastoideus was observed distally. As platysma myoides extended across the face, its fibers inserted onto **zygomaticus major** and blended with **depressor anguli oris**. The mid-face was a blend of multidirectional fibers, but after careful cleaning and dissection a pattern of delicate bundles was revealed. **Zygomaticus major** arose from the lateral-most portion of **orbicularis oculi** and the lateral zygomatic process, its body spread wider than usual. **Zygomaticus minor** was normal in size and originated from the fibers around the mouth, including **orbicularis oris** and **levator labii superioris alaeque nasi,** and inserted onto the medial zygomatic bone and partially onto orbicularis oculi. A muscle that appeared to be **levator labii superioris alaeque nasi** on both sides originated from the midpoint of the maxilla and spread inferolaterally, inserting onto orbicularis oris. A bundle of fibers matching the normal direction of **levator labii superioris** was found deep to levator labii superioris alaeque nasi, originating from the zygomatic bone and inserting onto orbicularis oris and blending deeply with levator labii superioris alaeque nasi. Immediately deep to levator labii superioris alaeque nasi was a distinct transverse bundle of fibers which appeared to correspond to the **nasalis** muscle and partially inserted onto levator labii superioris alaeque nasi. Upon division and reflection of orbicularis oris and zygomaticus major, a **levator anguli oris** muscle was found underneath with normal origin and insertion. Orbicularis oculi was complete, but it seemed to converge in the midline and expand laterally, paralleling to the two partially fused irises. No **procerus** or **corrugator supercilli** could be identified. **Depressor anguli oris, depressor labii inferioris,** and **mentalis** were all present but were at least partially fused. They were mainly distinguished by direction of fibers. These muscles lay superficial to the deep, inferior orbicularis oris fibers. **Auricularis posterior, auricularis superior,** and **auricularis anterior** were all present on the left side; however, auricularis anterior was not a distinct muscle on the right side. On the left side, auricularis superior and anterior were partially fused, and the fibers of auricularis anterior were directed towards the facial muscle orbicularis oculi and blended with it. Bilaterally, **digastricus anterior** had two heads instead of one (as is usually the case in karyotypically normal humans) and seemed very broad; its medial head was fused with its counterpart in the midline, while its lateral head had orthogonal fibers fused with the superficial **mylohyoideus**.

Regarding the other muscles of the branchial arches, when **stylohyoideus** was defined on both sides, we found that a portion of the muscle had already been removed, but the origin was intact and determined to be the normal styloid process cartilage. Bilaterally, the **digastricus posterior** bellies were unusually broad but seemed to insert normally onto the mastoid portion of the petrosal/temporal bone. Mylohyoideus lacked a clear median raphe, and a peculiar muscle lying beneath seemed to correspond to the **intermandibularis anterior** muscle of several non-human species (Diogo and Wood, 2011), having transverse fibers that spanned the two halves of the anterior mandible. The **tongue muscles** were normal, but the insertions of muscles onto the hyoid bone could not be observed due to the removal of the hyoid cartilage prior to our dissections. On both sides, **temporalis** inserted not only onto the coronoid process of the mandible but along the entire medial side of the ramus, reaching the angle of the mandible before it terminated. This anomalous attachment could be due to the early developmental stage of the fetus. **Pterygoideus lateralis** and **pterygoideus medialis** were mostly normal, the lateral pterygoid origin extending slightly further superior and posterior on the cranium. The **masseter** attachment points appeared normal.

The extraocular muscles **rectus medialis** and **obliquus superior** were absent bilaterally. **Rectus superior** was doubled and slightly wider than normal. **Rectus lateralis** on the right and left sides of the body had normal insertions, and **rectus inferior** was slightly thinner than normal. On the inferior portion of the eye, the **obliquus inferior** muscles on both sides of the body had anomalous configurations: each gave off one slip which fused with its counterpart on the other side and one slip which inserted onto the nearest lateral rectus. On the right, another slip inserted onto lateral rectus, while the left inferior oblique gave off a slip to insert into a supernumerary band of muscle parallel to the limbus of the eye.

2.7 Bones of the cranium (Figs. 2.15–2.17)

The skeletal structure of the skull is briefly summarized here to complement the myological information provided in the Sections above. The most unusual feature of the cranium was the single orbit. This feature appeared to suggest the presence of a single eye, but closer inspection revealed that the orbit contained not one, but two fused eyes with one iris and two pupils. The caruncle was located near the midline, inferomedial to the eyes, and the structure of the eyelid was unusual, particularly the upper portion which appeared to have two lids with a medial portion (Fig. 2.15). The frontal consisted of a single bone, missing the metopic suture. However, the CT scan showed two presumed ossification centers on the frontal bone slightly lateral to the midline but fused at the midline to form the single frontal bone. There appeared to be a single optic canal through the sphenoid bone. The maxilla was markedly smaller than normal and had a midline protuberance that served as the attachment for levator labii superioris alaeque nasi, inferior orbicularis oculi, and nasalis. The nasal bones were absent and, unlike in most cases of cyclopia, no proboscis or rudimentary nasal structures were observed. The ethmoid, inferior nasal concha, and vomer also seemed to be absent. The presence of the palatine bone could not be determined, as the mass of the maxilla may or may not have included part of a palatine bone. The mandible was curved inferiorly and slightly elongated.

2.8 Table 2—Muscular anomalies in 28-week Trisomy 18 cyclopic fetus compared with documented cases of Trisomies 18, 13, and 21

For each section of the table (Supernumerary, Absent and Variant muscles), each anomaly is independent of the others (e.g., an absent muscle is not a variant muscle, and they are treated separately in the population). In the Variant sections, a normal muscle in the sample population does not necessarily mean that it was present, but that it was non-variant in its presence or absence. Data were collected from the following papers: Aziz, 1979; 1980; 1981a,b; Barash et al., 1970; Bersu and Ramirez-Castro, 1977; Bersu, 1980; Colacino and Pettersen, 1978; Dunlap et al., 1986; Pettersen, 1979; Pettersen et al., 1979; Ramirez-Castro and Bersu, 1978; Urban and Bersu, 1987. B = Bilateral, R = Right, L = Left side of the body on individuals, when reported in literature. yrs = years.

Muscle anomalies in Trisomy 18 cyclopic fetus dissected by us and the respective side where these anomalies were found, in this fetus		Ratio of respective anomaly (presence versus absence, shown in bold), according to reports on Trisomy 13 by other authors	Ratio of respective anomaly (presence versus absence, shown in bold), according to reports on Trisomy 18 by other authors	Ratio of respective anomaly (presence versus absence, shown in bold), according to reports on Trisomy 21 by other authors
Supernumerary				
Presence of **musculus interosseus accessorius**	Bilateral	**0/24** (not in 10 male neonates, 6 neonates, male fetus, 5 female neonates, male 6 yrs, female fetus)	**0/26** (not in 5 male neonates, 5 neonates, 2 male fetuses, 12 female neonates, 2 female fetuses)	**0/7** (not in female fetus, 2 male fetuses, female 2.3 yrs, female 2.4 yrs, female 24 yrs, female 29 yrs)
Presence of **coracobrachialis profundus**	Left	**2** (female neonate L, female neonate L)/**24** (not in 10 male neonates, 6 neonates, male fetus, 3 female neonates, male 6 yrs, female fetus)	**0/26** (not in 5 male neonates, 5 neonates, 2 male fetuses, 12 female neonates, 2 female fetuses)	**0/7** (not in female fetus, 2 male fetuses, female 2.3 yrs, female 2.4 yrs, female 24 yrs, female 29 yrs)
Presence of separate fleshy belly of **fibularis tertius**	Right	**0/20** (normal in 8 male neonates, 6 neonates, male fetus, 3 female neonates, male 6 yrs, female fetus)	**0/17** (normal in 2 male neonates, 5 neonates, male fetus, 8 female neonates, female fetus)	**0/5** (normal in female fetus, male fetus, female 2.3 yrs, female 2.4 yrs, female 24 yrs)
Presence of muscle slip(s) from **flexor digitorum superficialis** and **flexor carpi radialis** to **flexor pollicis longus**	1 on Left 2 on Right	**3** (female neonate L, male neonate L, male neonate R)/**24** (not in 8 male neonates, 6 neonates, male fetus, 4 female neonates, male 6 yrs, female fetus)	**8** (female neonate L, female neonate B, female neonate B, female neonate L, male neonate R, male fetus L, male neonate L, female fetus B)/**26** (not in 3 male neonates, 5 neonates, male fetus, 8 female neonates, female fetus)	**1** (male fetus B)/**7** (not in female fetus, male fetus, female 2.3 yrs, female 2.4 yrs, female 24 yrs, female 29 yrs)

Table 2 contd....

Table 2 contd.

Muscle anomalies in Trisomy 18 cyclopic fetus dissected by us and the respective side where these anomalies were found, in this fetus		Ratio of respective anomaly (presence versus absence, shown in bold), according to reports on Trisomy 13 by other authors	Ratio of respective anomaly (presence versus absence, shown in bold), according to reports on Trisomy 18 by other authors	Ratio of respective anomaly (presence versus absence, shown in bold), according to reports on Trisomy 21 by other authors
Absent				
Extensor pollicis brevis	Right	**0/24** (not in 10 male neonates, 6 neonates, male fetus, 5 female neonates, male 6 yrs, female fetus)	**4** (female neonate R, male fetus B, female fetus L, female fetus L)/**26** (present in 5 male neonates, 5 neonates, male fetus, 11 female neonates)	**0/7** (present in female fetus, 2 male fetuses, female 2.3 yrs, female 2.4 yrs, female 24 yrs, female 29 yrs)
1st **Lumbricalis**	Bilateral	**3** (female neonate B, male neonate R, male fetus B)/**24** (present in 9 male neonates, 6 neonates, 4 female neonates, male 6 yrs, female fetus)	**4** (female neonate L, female neonate L, male fetus L, female fetus R)/**26** (present in 5 male neonates, 5 neonates, male fetus, 10 female neonates, female fetus)	**0/7** (present in female fetus, 2 male fetuses, female 2.3 yrs, female 2.4 yrs, female 24 yrs, female 29 yrs)
Palmaris brevis	Bilateral	**20** (male neonate B, male neonate B, male neonate B, female neonate B, male neonate B, male neonate B, female neonate B, female neonate B, female neonate B, male neonate B, male neonate B, female fetus B, male fetus B, neonate, neonate, neonate, female neonate B, male neonate B, male neonate B, male neonate B)/**24** (present in 3 neonates, male 6 yrs)	**15** (female neonate B, female neonate L, female neonate B, female neonate B, female neonate B, female neonate B, female neonate B, female neonate R, male neonate B, male neonate B, male fetus B, male neonate B, female fetus B, female neonate B, neonate, neonate)/**26** (present in 2 male neonates, 3 neonates, male fetus, 4 female neonates, female fetus)	**1** (male fetus L)/**7** (present in female fetus, male fetus, female 2.3 yrs, female 2.4 yrs, female 24 yrs, female 29 yrs)
Palmaris longus	Left	**20** (male neonate B, male neonate B, male neonate B, female neonate B, female neonate B, female neonate B, male neonate B, male neonate B, female neonate B, male neonate B, male	**23** (female neonate B, female neonate B, female neonate B, female neonate B, female neonate B, female neonate B, male neonate B, male neonate B, male fetus B, female fetus L, female neonate B, female neonate B, female	**6** (female 24 yrs B, female 2.4 yrs B, female 2.3 yrs B, male fetus B, female fetus B, female 29 yrs)/**7** (present in male fetus)

Table 2 contd....

Table 2 contd.

Muscle anomalies in Trisomy 18 cyclopic fetus dissected by us and the respective side where these anomalies were found, in this fetus		Ratio of respective anomaly (presence versus absence, shown in bold), according to reports on Trisomy 13 by other authors	Ratio of respective anomaly (presence versus absence, shown in bold), according to reports on Trisomy 18 by other authors	Ratio of respective anomaly (presence versus absence, shown in bold), according to reports on Trisomy 21 by other authors
Absent				
		neonate B, female fetus B, male fetus B, neonate, neonate, neonate, female neonate B, male neonate B, male neonate B, male neonate B)/**24** (present in 3 neonates, male 6 yrs)	neonate B, female neonate B, male neonate L, male neonate B, male fetus B, male neonate B, female fetus L, female neonate B, female neonate B, neonate, neonate)/**26** (present in 3 neonates)	
Opponens pollicis	Right	**1** (male fetus R)/**24** (present in 10 male neonates, 6 neonates, 5 female neonates, male 6 yrs, female fetus)	**6** (female neonate B, female neonate B, female neonate L, female neonate B, female neonate B, female neonate B)/**26** (present in 5 male neonates, 5 neonates, 2 male fetuses, 6 female neonates, 2 female fetuses)	**0/7** (present in female fetus, 2 male fetuses, female 2.3 yrs, female 2.4 yrs, female 24 yrs, female 29 yrs)
Abductor pollicis brevis	Right	**1** (female neonate L)/**24** (present in 10 male neonates, 6 neonates, male fetus, 4 female neonates, male 6 yrs, female fetus)	**9** (male neonate R, female neonate R, female neonate B, female neonate L, female neonate B, female neonate B, male neonate B, female fetus L, female neonate B)/**26** (present in 3 male neonates, 5 neonates, 2 male fetuses, 6 female neonates, female fetus)	**0/7** (present in female fetus, 2 male fetuses, female 2.3 yrs, female 2.4 yrs, female 24 yrs, female 29 yrs)
Flexor pollicis brevis	Right	**1** (female neonate L)/**24** (present in 10 male neonates, 6 neonates, male fetus, 4 female neonates, male 6 yrs, female fetus)	**8** (male neonate R, female neonate R, female neonate R, female neonate L, female neonate B, female neonate B, male neonate B, female neonate B)/**26** (present in 3 male neonates, 5 neonates, 2 male fetuses, 6 female neonates, 2 female fetuses)	**0/7** (present in female fetus, 2 male fetuses, female 2.3 yrs, female 2.4 yrs, female 24 yrs, female 29 yrs)

Table 2 contd....

Table 2 contd.

Muscle anomalies in Trisomy 18 cyclopic fetus dissected by us and the respective side where these anomalies were found, in this fetus		Ratio of respective anomaly (presence versus absence, shown in bold), according to reports on Trisomy 13 by other authors	Ratio of respective anomaly (presence versus absence, shown in bold), according to reports on Trisomy 18 by other authors	Ratio of respective anomaly (presence versus absence, shown in bold), according to reports on Trisomy 21 by other authors
Absent				
Popliteus	Right	**0/20** (normal in 8 male neonates, 6 neonates, male fetus, 3 female neonates, male 6 yrs, female fetus)	**0/17** (present in 2 male neonates, 5 neonates, male fetus, 8 female neonates, female fetus)	**0/5** (present in female fetus, male fetus, female 2.3 yrs, female 2.4 yrs, female 24 yrs)
Auricularis anterior	Right	**0/20** (present in 8 male neonates, 6 neonates, male fetus, 3 female neonates, male 6 yrs, female fetus)	**1** (female neonate B)/**17** (present in 2 male neonates, 5 neonates, male fetus, 7 female neonates, female fetus)	**0/5** (present in female fetus, male fetus, female 2.3 yrs, female 2.4 yrs, female 24 yrs)
Corrugator supercilli	Bilateral	**0/20** (present in 8 male neonates, 6 neonates, male fetus, 3 female neonates, male 6 yrs, female fetus)	**0/17** (present in 2 male neonates, 5 neonates, male fetus, 8 female neonates, female fetus)	**0/5** (present in female fetus, male fetus, female 2.3 yrs, female 2.4 yrs, female 24 yrs)
Procerus	Bilateral	**0/20** (present in 8 male neonates, 6 neonates, male fetus, 3 female neonates, male 6 yrs, female fetus)	**0/17** (present in 2 male neonates, 5 neonates, male fetus, 8 female neonates, female fetus)	**0/5** (present in female fetus, male fetus, female 2.3 yrs, female 2.4 yrs, female 24 yrs)
Obliquus superior	Bilateral	None documented	None documented	None documented
Rectus medialis	Bilateral	None documented	None documented	None documented
Variant				
Rhomboideus major and **rhomboideus minor** attach higher onto scapula than is usually seen	Bilateral	**0/24** (normal in 10 male neonates, 6 neonates, male fetus, 5 female neonates, male 6 yrs, female fetus)	**0/26** (normal in 5 male neonates, 5 neonates, 2 male fetuses, 12 female neonates, 2 female fetuses)	**0/7** (normal in female fetus, 2 male fetuses, female 2.3 yrs, female 2.4 yrs, female 24 yrs, female 29 yrs)
Rhomboids major and minor are fused	Bilateral	**0/24** (not in 10 male neonates, 6 neonates, male fetus, 5 female neonates, male 6 yrs, female fetus)	**2** (major and minor fused in female neonate B, female neonate B)/**26** (not in 5 male neonates, 5 neonates, 2 male fetuses, 10 female neonates, 2 female fetuses)	**0/7** (not in female fetus, 2 male fetuses, female 2.3 yrs, female 2.4 yrs, female 24 yrs, female 29 yrs)

Table 2 contd....

Table 2 contd.

Muscle anomalies in Trisomy 18 cyclopic fetus dissected by us and the respective side where these anomalies were found, in this fetus		Ratio of respective anomaly (presence versus absence, shown in bold), according to reports on Trisomy 13 by other authors	Ratio of respective anomaly (presence versus absence, shown in bold), according to reports on Trisomy 18 by other authors	Ratio of respective anomaly (presence versus absence, shown in bold), according to reports on Trisomy 21 by other authors
Variant				
Rhomboids major and minor are about 1/3 fibrous and 2/3 tendinous	Bilateral	**0/24** (normal in 10 male neonates, 6 neonates, male fetus, 5 female neonates, male 6 yrs, female fetus)	**0/26** (normal in 5 male neonates, 5 neonates, 2 male fetuses, 12 female neonates, 2 female fetuses)	**0/7** (normal in female fetus, 2 male fetuses, female 2.3 yrs, female 2.4 yrs, female 24 yrs, female 29 yrs)
Trapezius tightly associated/fused with **splenius capitis** at midline origin	Bilateral	**0/24** (not in 10 male neonates, 6 neonates, male fetus, 5 female neonates, male 6 yrs, female fetus)	**0/26** (not in 5 male neonates, 5 neonates, 2 male fetuses, 12 female neonates, 2 female fetuses)	**0/7** (not in female fetus, 2 male fetuses, female 2.3 yrs, female 2.4 yrs, female 24 yrs, female 29 yrs)
Latissimus dorsi and **teres major** fused toward humeral insertion	Bilateral	**0/24** (not in 10 male neonates, 6 neonates, male fetus, 5 female neonates, male 6 yrs, female fetus)	**0/26** (not in 5 male neonates, 5 neonates, 2 male fetuses, 12 female neonates, 2 female fetuses)	**0/7** (not in female fetus, 2 male fetuses, female 2.3 yrs, female 2.4 yrs, female 24 yrs, female 29 yrs)
Extra **biceps brachii** tendon inserting onto **pectoralis major**	Bilateral	**0/24** (not in 10 male neonates, 6 neonates, male fetus, 5 female neonates, male 6 yrs, female fetus)	9 (female neonate B, female neonate B, female neonate B, female neonate L, female neonate L, male neonate L, neonate, neonate, neonate)/**26** (not in 5 male neonates, 2 neonates, 2 male fetuses, 11 female neonates, 2 female fetuses)	**0/7** (not in female fetus, 2 male fetuses, female 2.3 yrs, female 2.4 yrs, female 24 yrs, female 29 yrs)
No bicipital aponeurosis	Bilateral	**0/24** (present in 10 male neonates, 6 neonates, male fetus, 5 female neonates, male 6 yrs, female fetus)	**0/26** (present in 5 male neonates, 5 neonates, 2 male fetuses, 12 female neonates, 2 female fetuses)	**0/7** (present in female fetus, 2 male fetuses, female 2.3 yrs, female 2.4 yrs, female 24 yrs, female 29 yrs)
No short head of biceps brachii	Left	2 (female neonate L, male fetus L)/**24** (present in 10 male neonates, 6 neonates, 4 female neonates, male 6 yrs, female fetus)	**0/26** (present in 5 male neonates, 5 neonates, 2 male fetuses, 12 female neonates, 2 female fetuses)	**0/7** (present in female fetus, 2 male fetuses, female 2.3 yrs, female 2.4 yrs, female 24 yrs, female 29 yrs)

Table 2 contd....

Table 2 contd.

Muscle anomalies in Trisomy 18 cyclopic fetus dissected by us and the respective side where these anomalies were found, in this fetus		Ratio of respective anomaly (presence versus absence, shown in bold), according to reports on Trisomy 13 by other authors	Ratio of respective anomaly (presence versus absence, shown in bold), according to reports on Trisomy 18 by other authors	Ratio of respective anomaly (presence versus absence, shown in bold), according to reports on Trisomy 21 by other authors
Variant				
Short head of biceps brachii fused with **coracobrachialis**	Right	**0/24** (not in 10 male neonates, 6 neonates, male fetus, 5 female neonates, male 6 yrs, female fetus)	**3** (female neonate B, female neonate R, female neonate B)/**26** (not in 5 male neonates, 9 female neonates, 5 neonates, 2 male fetuses, 2 female fetuses)	**1** (accessory heads in female 29 yrs L)/**7** (not in female 24 yrs, female 2.4 yrs, female 2.3 yrs, male fetus, female fetus)
Coracobrachialis attaches all along the anteromedial proximal humerus	Right	**0/24** (normal in 10 male neonates, 6 neonates, male fetus, 5 female neonates, male 6 yrs, female fetus)	**0/26** (normal in 5 male neonates, 5 neonates, 2 male fetuses, 12 female neonates, 2 female fetuses)	**0/7** (normal in female fetus, 2 male fetuses, female 2.3 yrs, female 2.4 yrs, female 24 yrs, female 29 yrs)
Pectoralis major fused with **deltoideus**	Bilateral	**0/24** (not in 10 male neonates, 6 neonates, male fetus, 5 female neonates, male 6 yrs, female fetus)	**21** (female neonate B, female neonate B, female neonate B, female neonate B, female neonate B, female neonate B, male neonate B, male neonate B, female neonate B, female neonate B, female neonate B, female neonate B, male neonate B, male neonate B, male fetus B, female fetus B, female neonate B, female neonate L, neonate, neonate, neonate B)/**26** (not in male neonate, 2 neonates, male fetus, female fetus)	**0/7** (not in female fetus, 2 male fetuses, female 2.3 yrs, female 2.4 yrs, female 24 yrs, female 29 yrs)
Extensor pollicis brevis is very small/diminutive and fused with **abductor pollicis longus**	Left	**0/24** (normal in 10 male neonates, 6 neonates, male fetus, 5 female neonates, male 6 yrs, female fetus)	**0/26** (normal in 5 male neonates, 5 neonates, 2 male fetuses, 12 female neonates, 2 female fetuses)	**0/7** (normal in female fetus, 2 male fetuses, female 2.3 yrs, female 2.4 yrs, female 24 yrs, female 29 yrs)
Some tendons of the abductor pollicis longus blend with the fibers of **abductor pollicis brevis**	Left	**0/24** (not in 10 male neonates, 6 neonates, male fetus, 5 female neonates, male 6 yrs, female fetus)	**0/26** (not in 5 male neonates, 5 neonates, 2 male fetuses, 12 female neonates, 2 female fetuses)	**0/7** (not in female fetus, 2 male fetuses, female 2.3 yrs, female 2.4 yrs, female 24 yrs, female 29 yrs)

Table 2 contd....

Table 2 contd.

Muscle anomalies in Trisomy 18 cyclopic fetus dissected by us and the respective side where these anomalies were found, in this fetus		Ratio of respective anomaly (presence versus absence, shown in bold), according to reports on Trisomy 13 by other authors	Ratio of respective anomaly (presence versus absence, shown in bold), according to reports on Trisomy 18 by other authors	Ratio of respective anomaly (presence versus absence, shown in bold), according to reports on Trisomy 21 by other authors
Variant				
Abductor pollicis longus tendons tripled and two insert onto the above abductor pollicis brevis	Left	**0/24** (not in 10 male neonates, 6 neonates, male fetus, 5 female neonates, male 6 yrs, female fetus)	**0/26** (not in 5 male neonates, 5 neonates, 2 male fetuses, 12 female neonates, 2 female fetuses)	**0/7** (not in female fetus, 2 male fetuses, female 2.3 yrs, female 2.4 yrs, female 24 yrs, female 29 yrs)
Extensor carpi radialis longus and **extensor carpi radialis brevis** fused	Bilateral	**1** (female neonate R)/**24** (not in 10 male neonates, male 6 yrs, 6 neonates, male fetus, 4 female neonates, female fetus)	**4** (female neonate B, female neonate L, male neonate B, male fetus L)/**26** (not in 4 male neonates, 5 neonates, male fetus, 10 female neonates, 2 female fetuses)	**0/7** (not in female fetus, 2 male fetuses, female 2.3 yrs, female 2.4 yrs, female 24 yrs, female 29 yrs)
Tendon of **extensor digiti minimi** attaches also onto digit 4, but it does not seem to go from distal portion of tendon to digit 5	Left	**0/24** (normal in 10 male neonates, 6 neonates, male fetus, 5 female neonates, male 6 yrs, female fetus)	**0/26** (normal in 5 male neonates, 5 neonates, 2 male fetuses, 12 female neonates, 2 female fetuses)	**0/7** (normal in female fetus, 2 male fetuses, female 2.3 yrs, female 2.4 yrs, female 24 yrs, female 29 yrs)
No **extensor digitorum** tendon going to digit 5	Bilateral	**0/24** (present in 10 male neonates, 6 neonates, male fetus, 5 female neonates, male 6 yrs, female fetus)	**0/26** (present in 5 male neonates, 5 neonates, 2 male fetuses, 12 female neonates, 2 female fetuses)	**0/7** (present in female fetus, 2 male fetuses, female 2.3 yrs, female 2.4 yrs, female 24 yrs, female 29 yrs)
No **flexor digitorum superficialis** tendon to digit 5	Bilateral	**7** (male neonate B, female neonate R, female neonate L, male fetus L, female neonate B, male neonate B, male 6 yrs R, hypoplastic on L)/**24** (present in 8 male neonates, 6 neonates, 2 female neonates, female fetus)	**9** (female neonate L, male neonate B, female neonate B, male neonate B, female neonate B, female fetus L, female neonate B, female neonate R, male neonate B)/**26** (present in 3 male neonates, 5 neonates, 2 male fetuses, 6 female neonates, female fetus)	**3** (female fetus L, female 24 yrs R, male fetus B)/**7** (present in male fetus, female 2.3 yrs, female 2.4 yrs, female 29 yrs)
Extensor carpi ulnaris, extensor digit minimi and extensor digitiorum on left side are fused proximally	Left	**0/24** (not in 10 male neonates, 6 neonates, male fetus, 5 female neonates, male 6 yrs, female fetus)	**0/26** (not in 5 male neonates, 5 neonates, 2 male fetuses, 12 female neonates, 2 female fetuses)	**0/7** (not in female fetus, 2 male fetuses, female 2.3 yrs, female 2.4 yrs, female 24 yrs, female 29 yrs)

Table 2 contd....

Table 2 contd.

Muscle anomalies in Trisomy 18 cyclopic fetus dissected by us and the respective side where these anomalies were found, in this fetus		Ratio of respective anomaly (presence versus absence, shown in bold), according to reports on Trisomy 13 by other authors	Ratio of respective anomaly (presence versus absence, shown in bold), according to reports on Trisomy 18 by other authors	Ratio of respective anomaly (presence versus absence, shown in bold), according to reports on Trisomy 21 by other authors
Variant				
Flexor carpi radialis and **pronator teres** fused	Bilateral	**2** (male neonate R, female neonate B)/**24** (not in 9 male neonates, 6 neonates, male fetus, 4 female neonates, male 6 yrs, female fetus)	**0/26** (not in 5 male neonates, 5 neonates, 2 male fetuses, 12 female neonates, 2 female fetuses)	**0/7** (not in female fetus, 2 male fetuses, female 2.3 yrs, female 2.4 yrs, female 24 yrs, female 29 yrs)
Opponens pollicis, flexor pollicis and **abductor pollicis brevis** all fused	Left	**0/24** (not in 10 male neonates, 6 neonates, male fetus, 5 female neonates, male 6 yrs, female fetus)	**0/26** (not in 5 male neonates, 5 neonates, 2 male fetuses, 12 female neonates, 2 female fetuses)	**0/7** (not in female fetus, 2 male fetuses, female 2.3 yrs, female 2.4 yrs, female 24 yrs, female 29 yrs)
The oblique head of the **adductor pollicis** is separated into distinct bundles	Left	**0/24** (normal in 10 male neonates, 6 neonates, male fetus, 5 female neonates, male 6 yrs, female fetus)	**0/26** (normal in 5 male neonates, 5 neonates, 2 male fetuses, 12 female neonates, 2 female fetuses)	**0/7** (normal in female fetus, 2 male fetuses, female 2.3 yrs, female 2.4 yrs, female 24 yrs, female 29 yrs)
Adductor pollicis accessorius present, going from metacarpal 1 to base of proximal phalanx 1	Bilateral	**5** (male neonate B, male neonate B, female neonate B, female neonate B, female neonate B, female neonate L)/**24** (not in 8 male neonates, 6 neonates, male fetus, 2 female neonates, male 6 yrs, female fetus)	**10** (female neonate B, female neonate B, female neonate R, female neonate B, female neonate B, male neonate B, male neonate B, male fetus B, male neonate B, female fetus B)/**26** (not in 2 male neonates, 5 neonates, male fetus, 7 female neonates, female fetus)	**2** (female 29 yrs B, male fetus B)/**7** (not in female fetus, male fetus, female 2.3 yrs, female 2.4 yrs, female 24 yrs)
Pronator teres has two bellies, superficial and deep	Bilateral	**2** (male neonate R, female neonate B)/**24** (normal in 9 male neonates, 6 neonates, male fetus, 4 female neonates, male 6 yrs, female fetus)	**0/26** (normal in 5 male neonates, 5 neonates, 2 male fetuses, 12 female neonates, 2 female fetuses)	**0/7** (normal in female fetus, 2 male fetuses, female 2.3 yrs, female 2.4 yrs, female 24 yrs, female 29 yrs)
Extensor carpi radialis longus and brevis fused with **brachioradialis**	Left	**0/24** (normal in 10 male neonates, 6 neonates, male fetus, 5 female neonates, male 6 yrs, female fetus)	**0/26** (normal in 5 male neonates, 5 neonates, 2 male fetuses, 12 female neonates, 2 female fetuses)	**0/7** (normal in female fetus, 2 male fetuses, female 2.3 yrs, female 2.4 yrs, female 24 yrs, female 29 yrs)

Table 2 contd....

Table 2 contd.

Muscle anomalies in Trisomy 18 cyclopic fetus dissected by us and the respective side where these anomalies were found, in this fetus		Ratio of respective anomaly (presence versus absence, shown in bold), according to reports on Trisomy 13 by other authors	Ratio of respective anomaly (presence versus absence, shown in bold), according to reports on Trisomy 18 by other authors	Ratio of respective anomaly (presence versus absence, shown in bold), according to reports on Trisomy 21 by other authors
Variant				
Origin of the flexor digitorum superficialis is only from the ulna and the common flexor tendon (no radial head)	Bilateral	**4** (female neonate R, female neonate R, male neonate B, male neonate B)/**24** (normal in 8 male neonates, 6 neonates, male fetus, 3 female neonates, male 6 yrs, female fetus)	**2** (male fetus R, female fetus L)/**26** (normal in 5 male neonates, 5 neonates, male fetus, 12 female neonates, female fetus)	**0/7** (normal in female fetus, 2 male fetuses, female 2.3 yrs, female 2.4 yrs, female 24 yrs, female 29 yrs)
Extensor pollicis longus tendon split in two, but these two structures fuse again with each other before insertion	Bilateral	**0/24** (normal in 10 male neonates, 6 neonates, male fetus, 5 female neonates, male 6 yrs, female fetus)	**0/26** (normal in 5 male neonates, 5 neonates, 2 male fetuses, 12 female neonates, 2 female fetuses)	**0/7** (normal in female fetus, 2 male fetuses, female 2.3 yrs, female 2.4 yrs, female 24 yrs, female 29 yrs)
Extensor indicis tendon doubled, going both to digits 2 and 3	Right	**1** (female neonate B)/**24** (not in male 6 yrs, 10 male neonates, 6 neonates, 4 female neonates, female fetus, male fetus)	**12** (female neonate B, female neonate L, female neonate R, male neonate B, female neonate B, male fetus R, female neonate B, female neonate B, male neonate B, male neonate B, male neonate B, female fetus B)/**26** (not in male neonate, 5 neonates, male fetus, 6 female neonates, female fetus)	**2** (female 29 yrs L, male fetus L)/**7** (not in female 24 yrs, female 2.4 yrs, female 2.3 yrs, female fetus)
Flexor pollicis longus has a bifurcated tendon going to the base of proximal phalange and to the distal phalanx of the thumb	Right	**2** (male neonate R, female neonate L)/**24** (not in 9 male neonates, 6 neonates, male fetus, 4 female neonates, male 6 yrs, female fetus)	**4** (female neonate B, female neonate B, male fetus L, male neonate R)/**26** (not in 4 male neonates, 5 neonates, 9 female neonates, male fetus, 2 female fetuses)	**0/7** (not in female fetus, 2 male fetuses, female 2.3 yrs, female 2.4 yrs, female 24 yrs, female 29 yrs)
Separate belly of flexor digitorum superficialis to digit 2, fused only proximally with the rest of the muscle	Right	**0/24** (normal in 10 male neonates, 6 neonates, male fetus, 5 female neonates, male 6 yrs, female fetus)	**0/26** (normal in 5 male neonates, 5 neonates, 2 male fetuses, 12 female neonates, 2 female fetuses)	**0/7** (normal in female fetus, 2 male fetuses, female 2.3 yrs, female 2.4 yrs, female 24 yrs, female 29 yrs)

Table 2 contd....

Table 2 contd.

Muscle anomalies in Trisomy 18 cyclopic fetus dissected by us and the respective side where these anomalies were found, in this fetus		Ratio of respective anomaly (presence versus absence, shown in bold), according to reports on Trisomy 13 by other authors	Ratio of respective anomaly (presence versus absence, shown in bold), according to reports on Trisomy 18 by other authors	Ratio of respective anomaly (presence versus absence, shown in bold), according to reports on Trisomy 21 by other authors
Variant				
Flexor digitorum superficialis sends slip to flexor digitorum profundus	Right	**3** (female neonate L, male neonate L, female neonate B)/**24** (not in 9 male neonates, 6 neonates, male fetus, 3 female neonates, male 6 yrs, female fetus)	**7** (female neonate B, female neonate R, female neonate L, female neonate B, male neonate B, male fetus L, male neonate R)/**26** (not in 3 male neonates, 5 neonates, male fetus, 8 female neonates, 2 female fetuses)	**2** (female fetus B, female 24 yrs B)/**7**(not in 2 male fetuses, female 2.3 yrs, female 2.4 yrs, female 29 yrs)
Fibularis tertius and **extensor digitorum longus** tendon to digit 4 exchange tendinous tissue	Right	**0/20** (normal in 8 male neonates, 6 neonates, male fetus, 3 female neonates, male 6 yrs, female fetus)	**0/17** (normal in 2 male neonates, 5 neonates, male fetus, 8 female neonates, female fetus)	**0/5** (normal in female fetus, male fetus, female 2.3 yrs, female 2.4 yrs, female 24 yrs)
Separate fleshy head of **fibularis tertius**	Right	**0/20** (not in 8 male neonates, 6 neonates, male fetus, 3 female neonates, male 6 yrs, female fetus)	**0/17** (not in 2 male neonates, 5 neonates, male fetus, 8 female neonates, female fetus)	**0/5** (not in female fetus, male fetus, female 2.3 yrs, female 2.4 yrs, female 24 yrs)
No **extensor digitorum longus** tendon to digit 5	Right	**0/20** (present in 8 male neonates, 6 neonates, male fetus, 3 female neonates, male 6 yrs, female fetus)	**0/17** (present in 2 male neonates, 5 neonates, male fetus, 8 female neonates, female fetus)	**0/5** (present in female fetus, male fetus, female 2.3 yrs, female 2.4 yrs, female 24 yrs)
Fibularis tertius insertion extends further distally	Right	**0/20** (normal in 8 male neonates, 6 neonates, male fetus, 3 female neonates, male 6 yrs, female fetus)	**0/17** (normal in 2 male neonates, 5 neonates, male fetus, 8 female neonates, female fetus)	**0/5** (normal in female fetus, male fetus, female 2.3 yrs, female 2.4 yrs, female 24 yrs)
Extensor digitorum brevis tendon to digit 2 sends tendons to digits 2 and 3	Right	**0/20** (not in 8 male neonates, 6 neonates, male fetus, 3 female neonates, male 6 yrs, female fetus)	**0/17** (not in 2 male neonates, 5 neonates, male fetus, 8 female neonates, female fetus)	**0/5** (not in female fetus, male fetus, female 2.3 yrs, female 2.4 yrs, female 24 yrs)
Gastrocnemius and **soleus** fused toward insertion into Achilles tendon	Bilateral	**0/20** (normal in 8 male neonates, 6 neonates, male fetus, 3 female neonates, male 6 yrs, female fetus)	**0/17** (normal in 2 male neonates, 5 neonates, male fetus, 8 female neonates, female fetus)	**0/5** (normal in female fetus, male fetus, female 2.3 yrs, female 2.4 yrs, female 24 yrs)

Table 2 contd....

Table 2 contd.

Muscle anomalies in Trisomy 18 cyclopic fetus dissected by us and the respective side where these anomalies were found, in this fetus		Ratio of respective anomaly (presence versus absence, shown in bold), according to reports on Trisomy 13 by other authors	Ratio of respective anomaly (presence versus absence, shown in bold), according to reports on Trisomy 18 by other authors	Ratio of respective anomaly (presence versus absence, shown in bold), according to reports on Trisomy 21 by other authors
Variant				
No **flexor digitorum brevis** tendon to digit 5	Bilateral	**5** (male neonate L, male neonate B, female fetus B, male fetus B, male 6 yrs L)/**20** (present in 6 male neonates, 6 neonates, 3 female neonates)	**0/17** (present in 2 male neonates, 5 neonates, male fetus, 8 female neonates, female fetus)	**0/5** (present in female fetus, male fetus, female 2.3 yrs, female 2.4 yrs, female 24 yrs)
No **adductor hallucis** transverse head	Left	**0/20** (present in 8 male neonates, 6 neonates, male fetus, 3 female neonates, male 6 yrs, female fetus)	**0/17** (present in 2 male neonates, 5 neonates, male fetus, 8 female neonates, female fetus)	**0/5** (present in female fetus, male fetus, female 2.3 yrs, female 2.4 yrs, female 24 yrs)
Tendon of flexor digitorum brevis to digit 4 was diminutive	Bilateral	**0/20** (normal in 8 male neonates, 6 neonates, male fetus, 3 female neonates, male 6 yrs, female fetus)	**0/17** (normal in 2 male neonates, 5 neonates, male fetus, 8 female neonates, female fetus)	**0/5** (normal in female fetus, male fetus, female 2.3 yrs, female 2.4 yrs, female 24 yrs)
Sternocleidomastoideus separated into fascicles	Bilateral	**0/20** (normal in 8 male neonates, 6 neonates, male fetus, 3 female neonates, male 6 yrs, female fetus)	**0/17** (normal in 2 male neonates, 5 neonates, male fetus, 8 female neonates, female fetus)	**0/5** (normal in female fetus, male fetus, female 2.3 yrs, female 2.4 yrs, female 24 yrs)
Sternocleidomastoideus very broad and forms a complex with the **trapezius** and **rhomboideus occipitalis**	Bilateral	**0/20** (normal in 8 male neonates, 6 neonates, male fetus, 3 female neonates, male 6 yrs, female fetus)	**0/17** (normal in 2 male neonates, 5 neonates, male fetus, 8 female neonates, female fetus)	**0/5** (normal in female fetus, male fetus, female 2.3 yrs, female 2.4 yrs, female 24 yrs)
Wide attachment for trapezius	Bilateral	**0/24** (normal in 10 male neonates, 6 neonates, male fetus, 5 female neonates, male 6 yrs, female fetus)	**0/26** (normal in 5 male neonates, 5 neonates, 2 male fetuses, 12 female neonates, 2 female fetuses)	**0/7** (normal in female fetus, 2 male fetuses, female 2.3 yrs, female 2.4 yrs, female 24 yrs, female 29 yrs)
Strong fascial connection between sternocleidomastoideus and **platysma myoides** distally	Bilateral	**0/20** (normal in 8 male neonates, 6 neonates, male fetus, 3 female neonates, male 6 yrs, female fetus)	**0/17** (normal in 2 male neonates, 5 neonates, male fetus, 8 female neonates, female fetus)	**0/5** (normal in female fetus, male fetus, female 2.3 yrs, female 2.4 yrs, female 24 yrs)

Table 2 contd....

Table 2 contd.

Muscle anomalies in Trisomy 18 cyclopic fetus dissected by us and the respective side where these anomalies were found, in this fetus		Ratio of respective anomaly (presence versus absence, shown in bold), according to reports on Trisomy 13 by other authors	Ratio of respective anomaly (presence versus absence, shown in bold), according to reports on Trisomy 18 by other authors	Ratio of respective anomaly (presence versus absence, shown in bold), according to reports on Trisomy 21 by other authors
Variant				
Platsyma myoides sends fibers into **zygomaticus major** and **depressor anguli oris**	Bilateral	**0/20** (normal in 8 male neonates, 6 neonates, male fetus, 3 female neonates, male 6 yrs, female fetus)	**0/17** (normal in 2 male neonates, 5 neonates, male fetus, 8 female neonates, female fetus)	**0/5** (normal in female fetus, male fetus, female 2.3 yrs, female 2.4 yrs, female 24 yrs)
Inferior part of **orbicularis oris** only consists of deep portion	Bilateral	**0/20** (normal in 8 male neonates, 6 neonates, male fetus, 3 female neonates, male 6 yrs, female fetus)	**0/17** (normal in 2 male neonates, 5 neonates, male fetus, 8 female neonates, female fetus)	**0/5** (normal in female fetus, male fetus, female 2.3 yrs, female 2.4 yrs, female 24 yrs)
Digastricus anterior broad and with two heads; lateral head fused to superficial mylohyoid; medial head fused to counterpart	Bilateral	**4** (female neonate B, male neonate L, female fetus B, male neonate B)/**20** (not in 6 male neonates, 6 neonates, male fetus, 2 female neonates, male 6 yrs)	**5** (male fetus B, female neonate B, female neonate B, neonate B, female neonate R)/**17** (not in 2 male neonates, 4 neonates, 5 female neonates, female fetus)	**0/5** (not in female fetus, male fetus, female 2.3 yrs, female 2.4 yrs, female 24 yrs)
Digastricus posterior broad	Bilateral	**0/20** (normal in 8 male neonates, 6 neonates, male fetus, 3 female neonates, male 6 yrs, female fetus)	**0/17** (normal in 2 male neonates, 5 neonates, male fetus, 8 female neonates, female fetus)	**0/5** (normal in female fetus, male fetus, female 2.3 yrs, female 2.4 yrs, female 24 yrs)
Mylohyoideus lacks medial raphe and is separated into deep and superficial portions	Bilateral	**1** (lacked median raphe and undirected fibers B and lacked posterior portion on R in female neonate)/**20** (normal in 8 male neonates, 6 neonates, male fetus, 2 female neonates, male 6 yrs, female fetus)	**1** (female neonate B)/**17** (normal in 3 male neonates, 5 neonates, male fetus, 6 female neonates, female fetus)	**0/5** (normal in female fetus, male fetus, female 2.3 yrs, female 2.4 yrs, female 24 yrs)
Rectus superior slightly doubled	Bilateral	None documented	None documented	None documented
Both **obliquus inferior** muscles gave off one slip to fuse to each other, one slip to insert onto their nearest lateral recti. On the right another slip inserted onto the lateral rectus, while left inferior oblique gave off a slip to insert into a supernumerary band of muscle which was parallel to the limbus of the eye	Bilateral	None documented	None documented	None documented

CHAPTER 3

Comparative Anatomy of Muscular Anomalies in Trisomies 13, 18, and 21

3.1 Introduction

In this Chapter we summarize an extensive literature review on muscle defects in Trisomies 13, 18, and 21, which are categorized in Table 3 and shown in Figs. 3.1–3.73 (references for this Chapter are listed in the table and figure captions). Although Trisomies 18, 13, and 21 are associated with the presence of three different chromosomes, our data clearly show repeated patterns of muscle defects in at least some individuals with all three conditions. These patterns help to identify the phenotypes that characterize particular trisomies, and, more importantly, elucidate commonalities between them. These commonalities may reveal conservation and constraint of body form even in severe, seemingly chaotic, developmental defects. However, Trisomy 21 produces fewer anatomical defects than Trisomies 13 and 18, as might have been predicted based on the longer life expectancy of individuals with Down syndrome (60+ years) versus those with Trisomies 13 and 18 (> 1 year; see Chapter 1).

3.2 Head and neck (Figs. 3.1–3.26)

Platysma cervicale was the only reported supernumerary muscle of facial expression. It was reported in five (25%) of 20 Trisomy 13 individuals, 13 (76.5%) of 17 Trisomy 18 individuals, and five (100%) of five Trisomy 21 individuals. Counting each side of the body separately, the platysma cervicale was present in seven (17.5%) of 40 Trisomy 13 sides, 26 (76.5%) of 34 Trisomy 18 sides, and 10 (100%) of 10 Trisomy 21 sides. Presence of this muscle was unexpectedly prevalent, especially in Trisomies 18 and 21. Both the platysma myoides and platysma cervicale appear early in normal human development, but, contrary, to the platysma myoides, the platysma cervicale disappears in early (embryonic) developmental stages. The presence of the platysma cervicale at later stages of development in trisomic individuals, including adults with Down syndrome, may therefore be related to a developmental delay and constitute a true case of evolutionary atavism (see Chapter 6).

The **digastricus anterior** was variant in four (20%) of 20 Trisomy 13 individuals and five (29.4%) of 17 Trisomy 18 individuals; per side of the body, this muscle was variant in seven (17.5%) of 40 Trisomy 13 sides and nine (26.4%) of 34 Trisomy 18 sides. The **digastricus posterior** was variant in five (25%) of 20 Trisomy 13 individuals and four (24%) of 17 Trisomy 18 individuals; per side of the body this muscle was variant in 10 (25%) of 40 Trisomy 13 sides and six (18%) of 34 Trisomy 18 sides. The supernumerary **mentohyoideus** muscle (e.g., Fig. 3.6) was reported in four (24%) of 17 Trisomy 18 individuals; per side of the body, this muscle was reported in eight (20%) of 40 Trisomy 18 sides. The **stylohyoideus** was absent in five (25%) of 20 Trisomy 13 individuals and 7 (41.2%) of 17 Trisomy 18 individuals; per side of the body this muscle was absent in eight (20%) of 40 Trisomy 13 sides, and 10 (29.4%) of 34 Trisomy 18 sides. Stylohyoideus was variant in two (10%) of 20 Trisomy 13 individuals and 5 (29.4%) of 17 Trisomy 18 individuals; per side of the body this muscle was variant in four (10%) of 40 Trisomy 13 sides, and seven (20.6%) of 34 Trisomy 18 sides. **Omohyoideus** was variant in seven (29.1%) of 24 Trisomy 13 individuals, and 11 (42.3%) of 26 Trisomy 18 individuals (there were not enough data reported on the specific sides of the bodies to report for this muscle).

3.3 Back and pectoral region (Figs. 3.27–3.38)

A prevalent muscle found in Trisomies 13 and 18 but not in karyotypically normal humans is the **rhomboideus occipitalis**, reported in one (4.2%) of 24 Trisomy 13, and 8 (30.8%) of 26 Trisomy 18 individuals; per side of the body this muscle was present in two (4%) of 48 Trisomy 13 sides and 14 (27%) of 52 Trisomy 18 sides. In human triploidy (three sets of chromosomes) Moen et al. (1984) reported an interesting absent middle portion of the trapezius in all three cases studied (Fig. 3.27). The **"pectorodorsalis"** was reported in 16 (66.7%) of 24 Trisomy 13 individuals; per limb this muscle was present in 31 (64.5%) of 48 Trisomy 13 limbs. **"Subclavius posticus"** was reported in three (15%) of 20 Trisomy 13 individuals and five (29.4%) of 17 Trisomy 18 individuals; per limb this muscle was present in four (10%) of 40 Trisomy 13 limbs and seven (20.5%) of 34 Trisomy 18 limbs. **Sternalis** was reported in one (5.8%) of 17 Trisomy 18 individuals; per limb this muscle was present in two (5.8%) of 34 Trisomy 18 limbs. **"Tensor semivaginae scapulohumeralis"** was reported in one (4.2%) of 24 Trisomy 13 individuals and 10 (38.5%) of 26 Trisomy 18 individuals; per limb, this muscle was present in one (2%) of 48 Trisomy 13 limbs and 18 (34.6%) of 52 Trisomy 18 limbs. **Coracobrachialis profundus** was reported in two (10%) of 20 Trisomy 13 individuals and five (29.4%) of 17 Trisomy 18 individuals; per limb this structure was found in four (10%) of 40 Trisomy 13 limbs and seven (20.6%) of 34 Trisomy 18 limbs. **"Pectoralis minimus"** was reported in eight (30.8%) of 26 Trisomy 18 individuals; per limb this muscle was present in 12 (23%) of 52 Trisomy 18 limbs. **"Chondroepitrochlearis"** was reported in five (21%) of 24 Trisomy 13 individuals (there were not enough data given to give a per limb count on this muscle). **Pectoralis major** was variant in five (21%) of 24 Trisomy 13 individuals, and seven (27%) of 26 Trisomy 18 individuals; per limb this muscle was variant in nine (19%) of 48 Trisomy 13 limbs and 12 (23%) of 52 Trisomy 18 limbs. A fused deltopectoral complex was reported in 21 (81%) of 26 Trisomy 18 individuals; per limb this complex was reported in 41 (79%) of 52 Trisomy 18 limbs (see Fig. 2.1) **Pectoralis minor** was variant in ten (38.5%) of 26 Trisomy 18 individuals; per limb this muscle was variant in 13 (25%) of 52 Trisomy 18 limbs. **Subclavius** was absent in ten (50%) of 20 Trisomy 13 individuals, 11 (64.7%) of 17 Trisomy 18 individuals; per limb this muscle was absent in 13 (33%) of 40 Trisomy 13 sides, and 16 (47%) of 34 Trisomy 18 sides.

3.4 Upper limb (Figs. 3.37, 3.39–3.62)

Biceps brachii was variant in 12 (50%) of 24 Trisomy 13 individuals, 15 (58%) of 26 Trisomy 18 individuals and one (14%) of seven Trisomy 21 individuals; per limb this muscle was variant in 19 (40%) of 48 Trisomy 13 limbs, 26 (50%) of 52 Trisomy 18 limbs, and one (7%) of 14 Trisomy 21 limbs. **Triceps brachii** was variant in two (8%) of 24 Trisomy 13 individuals and 18 (69%) of 26 Trisomy 18 individuals; per limb this muscle was variant in three (6%) of 48 Trisomy 13 limbs and 36 (69%) of 52 Trisomy 18 limbs. **Extensor carpi radialis accessorius** was reported in five (19.2%) of 26 Trisomy 18 individuals; per limb this muscle was present in eight (15.4%) of 52 Trisomy 18 limbs. **Extensor indicis** was variant in seven (29%) of 24 Trisomy 13 individuals, 12 (46%) of 26 Trisomy 18 individuals and two (29%) of seven Trisomy 21 individuals; per limb this muscle was variant in 13 (27%) of 48 Trisomy 13 limbs, 21 (40.4%) of 52 Trisomy 18 limbs, and two (14%) of 14 Trisomy 21 limbs.

Ramirez-Castro and Bersu (1978) reported that, in their case 5, the first metacarpal from digit one was absent; the **abductor pollicis longus** inserted onto the trapezium, the **extensor pollicis longus** inserted onto the base of metacarpal two and the **extensor pollicis brevis** inserted onto the styloid process of the radius; in their case 2, there was abnormal attachments present for the tendons of the thumb muscles. In our overall literature review, the **flexor carpi radialis brevis** was reported in one (3.8%) of 26 Trisomy 18 individuals; per limb this muscle was present in two (3.8%) of 52 Trisomy 18 limbs. **Flexor carpi ulnaris accessorius** was reported in nine (34.6%) of 26 Trisomy 18 individuals; per limb this muscle was present in 13 (25%) of 52 Trisomy 18 limbs. **Radiocarpus** muscle was reported in five (20.8%) of 24 Trisomy 13 individuals and five (19.2%) of 26 Trisomy 18 individuals; per limb this muscle was reported in seven (14.6%) of 48 Trisomy 13 limbs, and 6 (11.5%) of 52 Trisomy 18 limbs. The **radiocarpus accessorius** was reported in two (8.3%) of 24 Trisomy 13 individuals; per limb this muscle was reported in two (4.2%) of 48 Trisomy 13 limbs. The **flexor pollicis longus accessorius** was reported in one (4.2%) of 24 Trisomy 13 individuals; per limb this muscle was reported in two (4.2%) of 48 Trisomy 13 limbs. **Extensor digitorum** was variant in one (4%) of 24 Trisomy 13 individuals and nine (35%) of 26 Trisomy 18 individuals; per limb this muscle was variant in one (2%) of 48 Trisomy 13 limbs and 18 (35%) of 52 Trisomy 18 limbs. **Palmaris longus** was absent in 20 (83%) of 24 Trisomy 13 individuals, 23 (88.5%) of 26 Trisomy 18 individuals and six (85.7%) of seven Trisomy 21 individuals; per limb this muscle was absent in 40 (83%) of 48 Trisomy 13 limbs, 43 (83%) of 52 Trisomy 18 limbs, and 12 (86%) of 14 Trisomy 21 limbs. **Flexor pollicis longus** was variant in two (8%) of 24 Trisomy 13 individuals, 13 (50%) of 26 Trisomy 18 individuals and one (14.3%) of seven Trisomy 21 individuals; per limb this muscle was variant in two (4.2%) of 48 Trisomy 13 limbs, 19 (37%) of 52 Trisomy 18 limbs, and two (14%) of 14 Trisomy 21 limbs. **Flexor pollicis brevis** was variant in two (8%) of 24 Trisomy 13 individuals and 23 (38.5%) of 26 Trisomy 18 individuals. Per limb, the flexor pollicis brevis was variant in three (6.3%) of 48 Trisomy 13 limbs and 15 (29%) of 52 Trisomy 18 limbs. **Flexor digitorum superficialis** was variant in six (25%) of 24 Trisomy 13 individuals, 11 (42.3%) of 26 Trisomy 18 individuals and two (29%) of seven Trisomy 21 individuals; per limb this muscle was variant in eight (16.6%) of 48 Trisomy 13 limbs, 18 (35%) of 52 Trisomy 18 limbs, and three (21.4%) of 14 Trisomy 21 limbs. **Extensor carpi radialis longus** and **extensor carpi radialis brevis** were variant in three (12.5%) of 24 Trisomy 13 individuals, and 10 (38.5%) of 26 Trisomy 18 individuals; per limb these muscles were variant in six (12.5%) of 48 Trisomy 13 limbs and 19 (37%) of 52 Trisomy 18 limbs. **Abductor pollicis longus** was variant in four (17%) of 24 Trisomy 13 individuals and nine (35%) of 26 Trisomy 18 individuals; per limb this muscle was variant in seven (15%) of 48 Trisomy 13 limbs, 14 (27%) of 52 Trisomy 18 limbs, and two (14%) of 14 Trisomy 21 limbs. **Extensor pollicis longus** and **extensor pollicis brevis** were variant in two (8%) of 24 Trisomy 13 individuals, 13 (50%) of 26 Trisomy 18 individuals and one (14.3%) of seven Trisomy 21 individuals; per limb these muscles are variant in three (6.3%) of 48 Trisomy 13 limbs, 19 (36.5%) of 52 Trisomy 18 limbs, and one (7%) of 14 Trisomy 21 limbs. **Flexor digitorum profundus** was variant in five (21%) of 24 Trisomy 13 individuals and six (23%) of 26 Trisomy 18 individuals; per limb this muscle was variant in six (12.5%) of 48 Trisomy 13 limbs and nine (17.3%) of 52 Trisomy 18 limbs. **Extensor digiti minimi** was variant in two (8%) of 24 Trisomy 13 individuals, 12 (46.2%) of 26 Trisomy 18 individuals and one (14.3%) of seven Trisomy 21 individuals; per limb this muscle was variant in three (6.3%) of 48 Trisomy 13 limbs, 22 (42.3%) of 52 Trisomy 18 limbs, and two (14%) of 14 Trisomy 21 limbs.

The second muscle of the **contrahentes** series (going to digit 2; usually humans only have one contrahens, which is the only going to digit 1 and named adductor pollicis in human anatomy) was reported in four (15.4%) of 26 Trisomy 18 individuals; per limb the contrahens II was reported in six (11.5%) of 52 Trisomy 18 limbs. **Palmaris brevis** was present, but variant in one (4.2%) of 24 Trisomy 18 individuals; per limb this muscle was present, but variant in one (20.8%) of 48 Trisomy 18 limbs. Palmaris brevis was absent in 20 (83%) of 24 Trisomy 13 individuals, 15 (57.7%) of 26 Trisomy 18 individuals and one (14.3%) of seven Trisomy 21 individuals; per limb this muscle was absent in 40 (83%) of 48 Trisomy 13 limbs, 28 (54%) of 52 Trisomy 18 limbs, and one (7%) of 14 Trisomy 21 limbs. **Abductor pollicis brevis** was absent in one (4.2%) of 24 Trisomy 13 individuals and nine (34.6%) of 26 Trisomy 18 individuals; per limb this muscle was absent in one (2%) of 48 Trisomy 13 limbs, and 14 (27%) of 52 Trisomy 18 limbs. The **lumbricales** were variant in five (21%) of 24 Trisomy 13 individuals, nine (35%) of 26 Trisomy 18 individuals and one (14.3%) of seven Trisomy 21 individuals; per limb these muscles were variant in six (12.5%) of 48 Trisomy 13 limbs, 12 (23%) of 52 Trisomy 18 limbs, and one (7%) of 14 Trisomy 21 limbs.

3.5 Lower limb (Figs. 3.63–3.73)

Biceps femoris was variant in one (5%) of 20 Trisomy 13 individuals and one (5.8%) of 17 Trisomy 18 individuals; per limb this muscle was variant in one (2.5%) of 40 Trisomy 13 limbs, and one (2.9%) of 34 Trisomy 18 limbs. A supernumerary muscle slip between the biceps femoris and **semitendinosus** was reported in three (17.6%) of 17 Trisomy 18 individuals; per limb, this muscle slip was reported in three (8.8%) of 34 Trisomy 18 limbs. A supernumerary muscle slip between the sacrotuberous ligament and semitendinosus was reported in two (11.8%) of 17 Trisomy 18 individuals; per limb this muscle slip was reported in two (5.9%) of 34 Trisomy 18 limbs. **Tennuissimus** was reported in two (11.8%) of 17 Trisomy 18 individuals; per limb this muscle was reported in two (5.9%) of 34 Trisomy 18 limbs. **Fibularis (peroneus) quinti digiti** was reported in two (10%) of 20 Trisomy 13 individuals and three (17.6%) of 17 Trisomy 18 individuals (there were not enough data to give a per limb count for this muscle). **Fibularis quartus** was reported in three (15%) of 20 Trisomy 13 individuals and five (29.4%) of 17 Trisomy 18 individuals; per limb this muscle was reported in six (15%) of 40 Trisomy 13 limbs and six (17.6%) of 34 Trisomy 18 limbs. **Extensor primi internodii hallucis** was reported in one (5.9%) of 17 Trisomy 18 individuals; per limb this muscle was reported in two (5.9%) of 34 Trisomy 18 limbs. **Peroneocalcaneus medialis** was reported in three (17.6%) of 17 Trisomy 18 individuals; per limb this muscle was reported in six (17.6%) of 34 Trisomy 18 limbs. **Fibularis tertius** was completely absent (i.e., including its tendon, which is usually present in karyotypically normal humans) in 12 (60%) of 20 Trisomy 13 individuals; per limb this muscle was completely absent in 24 (60%) of 40 Trisomy 13 limbs.

3.6 Table 3—Muscular anomalies reported by other authors in Trisomies 18, 13 and 21

Data collected from the following papers: Aziz, 1979; 1980; 1981a,b; Barash et al., 1970; Bersu and Ramirez-Castro, 1977; Bersu, 1980; Colacino and Pettersen, 1978; Dunlap et al., 1986; Pettersen, 1979; 1979; Ramirez-Castro and Bersu, 1978; Urban and Bersu, 1987.

Muscle anomalies	Trisomy 13 (presence versus absence, shown in bold)	Trisomy 18 (presence versus absence, shown in bold)	Trisomy 21 (presence versus absence, shown in bold)
Supernumerary			
Platysma cervicale (mouth to nuchal region)	5 (male 6 yrs L, male neonate R, male neonate B, female neonate L, male neonate B)/20 (not in female fetus, male fetus, 6 neonates, 5 male neonates, 2 female neonates)	13 (male fetus B, female fetus B, male neonate B, male neonate B, female neonate B, female neonate B, female neonate B, female neonate B, female neonate B, female neonate B, neonate B, neonate B, neonate B)/17 (not in 2 neonates, 2 female neonates)	5 (female 24 yrs B, female 2.4 yrs B, female 2.3 yrs B, male fetus B, female fetus B)/5
Levator glandulae thyroidae (thyroid cartilage to isthmus of gland)	2 (female neonate R, male neonate L)/20 (not in 7 male neonates, 6 neonates, male fetus, 2 female neonates, male 6 yrs, female fetus)	0/17 (not in 2 male neonates, 5 neonates, male fetus, 8 female neonates, female fetus)	0/5 (not in female fetus, male fetus, female 2.3 yrs, female 2.4 yrs, female 24 yrs)
Levator claviculae (between basilar portion of occipital bone and clavicle)	0/20 (not in 8 male neonates, 6 neonates, male fetus, 3 female neonates, male 6 yrs, female fetus)	0/17 (not in 2 male neonates, 5 neonates, male fetus, 8 female neonates, female fetus)	1 (female 24 yrs B)/5 (not in female 2.4 yrs, female 2.3 yrs, male fetus, female fetus)
Occipito-scapularis (superior nuchal line to serratus anterior)	0/24 (not in 10 male neonates, 6 neonates, male fetus, 5 female neonates, male 6 yrs, female fetus)	0/17 (not in 2 male neonates, 5 neonates, male fetus, 8 female neonates, female fetus)	2 (female 24 yrs L, male fetus L)/5 (not in female 2.4 yrs, female 2.3 yrs, female fetus)
Petropharyngeus (accessory levator muscle of pharynx, from undersurface of petrous portion of temporal bone to pharynx)	0/20 (not in 8 male neonates, 6 neonates, male fetus, 3 female neonates, male 6 yrs, female fetus)	0/17 (not in 2 male neonates, 5 neonates, male fetus, 8 female neonates, female fetus)	1 (female 2.3 yrs B)/5 (not in female 2.4 yrs, female 24 yrs, male fetus, female fetus)
Supracostalis (from first rib to third or fourth rib)	2 (male 6 yrs, male neonate L)/20 (not in 7 male neonates, 6 neonates, male fetus, 3 female neonates, female fetus)	0/17 (not in 2 male neonates, 5 neonates, male fetus, 8 female neonates, female fetus)	1 (female 2.3 yrs L)/5 (not in female 2.4 yrs, female 24 yrs, male fetus, female fetus)
Sternoclavicularis (from anterior surface of manubrium and capsule of sternoclavicular joint, to anterior surface of clavicle	0/20 (not in 8 male neonates, 6 neonates, male fetus, 3 female neonates, male 6 yrs, female fetus)	1 (female neonate R)/17 (not in 2 male neonates, 5 neonates, male fetus, 7 female neonates, female fetus)	0/5 (not in female fetus, male fetus, female 2.3 yrs, female 2.4 yrs, female 24 yrs)

Table 3 contd....

Table 3 contd.

Muscle anomalies	Trisomy 13 (presence versus absence, shown in bold)	Trisomy 18 (presence versus absence, shown in bold)	Trisomy 21 (presence versus absence, shown in bold)
Supernumerary			
Sternohyoideus azygos	**1** (neonate)/**20** (not in 8 male neonates, 5 neonates, male fetus, 3 female neonates, male 6 yrs, female fetus)	**2** (female neonate B, female neonate)/**17** (not in 2 male neonates, 5 neonates, male fetus, 6 female neonates, female fetus)	**0/5** (not in female fetus, male fetus, female 2.3 yrs, female 2.4 yrs, female 24 yrs)
Extra stylopharyngeus	**1** (male 6 yrs L)/**20** (not in 8 male neonates, 6 neonates, male fetus, 3 female neonates, female fetus)	**0/17** (not in 2 male neonates, 5 neonates, male fetus, 8 female neonates, female fetus)	**0/5** (not in female fetus, male fetus, female 2.3 yrs, female 2.4 yrs, female 24 yrs)
Rhomboideus occipitalis (from superior angle of scapula to occipital region of skull)	**1** (female neonate B)/**24** (not in 10 male neonates, 6 neonates, male fetus, 4 female neonates, male 6 yrs, female fetus)	**8** (female neonate B, female neonate R, male neonate B, male neonate B, male neonate R, female fetus B, female neonate B, neonate B)/**26** (not in 2 male neonates, 4 neonates, 2 male fetuses, 9 female neonates, female fetus)	**0/7** (not in female fetus, 2 male fetuses, female 2.3 yrs, female 2.4 yrs, female 24 yrs, female 29 yrs)
Mentohyoideus (from hyoid to mandible)	**0/20** (not in 8 male neonates, 6 neonates, male fetus, 3 female neonates, male 6 yrs, female fetus)	**4** (neonate B, male neonate B, female neonate B, female neonate B)/**17** (not in male neonates, 4 neonates, male fetus, 6 female neonates, female fetus)	**0/5** (not in female fetus, male fetus, female 2.3 yrs, female 2.4 yrs, female 24 yrs)
Muscle belly between central tendon and pericardium	**1** (male neonate L)/**20** (not in 7 male neonates, 6 neonates, male fetus, 3 female neonates, male 6 yrs, female fetus)	**0/17** (not in 2 male neonates, 5 neonates, male fetus, 8 female neonates, female fetus)	**0/5** (not in female fetus, male fetus, female 2.3 yrs, female 2.4 yrs, female 24 yrs)
Muscle between central tendon and pulmonary vein	**2** (female neonate R, male neonate B)/**20** (not in 7 male neonates, 6 neonates, male fetus, 2 female neonates, male 6 yrs, female fetus)	**0/17** (not in 2 male neonates, 5 neonates, male fetus, 8 female neonates, female fetus)	**0/5** (not in female fetus, male fetus, female 2.3 yrs, female 2.4 yrs, female 24 yrs)
Latissimo-condyloideus (inferior border of latissimus dorsi to long head of triceps brachii)	**0/24** (not in 10 male neonates, 6 neonates, male fetus, 5 female neonates, male 6 yrs, female fetus)	**4** (female neonate L, female neonate R, neonate L, neonate L)/**26** (not in 5 male neonates, 3 neonates, 2 male fetuses, 10 female neonates, 2 female fetuses)	**0/7** (not in female fetus, 2 male fetuses, female 2.3 yrs, female 2.4 yrs, female 24 yrs, female 29 yrs)
Hepaticodiaphragmaticus (diaphragm to hilum of liver)	**1** (male neonate L)/**20** (not in 7 male neonates, 6 neonates, male fetus, 3 female neonates, male 6 yrs, female fetus)	**0/17** (not in 2 male neonates, 5 neonates, male fetus, 8 female neonates, female fetus)	**0/5** (not in female fetus, male fetus, female 2.3 yrs, female 2.4 yrs, female 24 yrs)
Subclavius posticus (from upper margin of scapula and transverse scapular ligament to superior side of first rib)	**3** (male neonate L, male neonate L, male fetus B)/**20** (not in 6 male neonates, 6 neonates, 3 female neonates, male 6 yrs, female fetus)	**5** (male neonate B, female neonate B, female neonate L, male neonate L, female neonate R)/**17** (not in 5 neonates, male fetus, 5 female neonates, female fetus)	**0/5** (not in female fetus, male fetus, female 2.3 yrs, female 2.4 yrs, female 24 yrs)

Table 3 contd....

Table 3 contd.

Muscle anomalies	Trisomy 13 (presence versus absence, shown in bold)	Trisomy 18 (presence versus absence, shown in bold)	Trisomy 21 (presence versus absence, shown in bold)
Supernumerary			
Sternalis (from infraclavicular region around the sternum to lower ribs/rectus abdominis)	0/**20** (not in 8 male neonates, 6 neonates, male fetus, 3 female neonates, male 6 yrs, female fetus)	1 (male neonate B)/**17** (not in male neonate, 5 neonates, male fetus, 8 female neonates, female fetus)	0/**5** (not in female fetus, male fetus, female 2.3 yrs, female 2.4 yrs, female 24 yrs)
Levator tendinis musculi latissimus dorsi (originates from coracoid process and inserts onto shoulder joint capsule to latissimus dorsi)	1 (male neonate L)/**24** (not in 9 male neonates, 6 neonates, male fetus, 5 female neonates, male 6 yrs, female fetus)	0/**26** (not in 5 male neonates, 5 neonates, 2 male fetuses, 12 female neonates, 2 female fetuses)	0/**7** (not in female fetus, 2 male fetuses, female 2.3 yrs, female 2.4 yrs, female 24 yrs, female 29 yrs)
Pectorodorsalis (from pectoralis major to humerus)	16 (male neonate B, male neonate B, female neonate B, female neonate B, female neonate B, male 6 yrs R, male neonate B, male neonate B, female fetus B, male fetus B, neonate B, neonate B, neonate B, female neonate B, male neonate B, male neonate B)/**24** (not in 4 male neonates, 3 neonates, female neonate)	0/**26** (not in 5 male neonates, 5 neonates, 2 male fetuses, 12 female neonates, 2 female fetuses)	0/**7** (not in female fetus, 2 male fetuses, female 2.3 yrs, female 2.4 yrs, female 24 yrs, female 29 yrs)
Belly between pectoralis major tendon and medial epicondyle of humerus	0/**24** (not in 10 male neonates, 6 neonates, male fetus, 5 female neonates, male 6 yrs, female fetus)	5 (male neonate L, female neonate B, female neonate B, female neonate L, female neonate L)/**26** (not in 4 male neonates, 5 neonates, 2 male fetuses, 8 female neonates, 2 female fetuses)	0/**7** (not in female fetus, 2 male fetuses, female 2.3 yrs, female 2.4 yrs, female 24 yrs, female 29 yrs)
Belly between pectoralis major tendon and biceps brachii	0/**24** (not in 10 male neonates, 6 neonates, male fetus, 5 female neonates, male 6 yrs, female fetus)	4 (male neonate L, female neonate B, female neonate B, female neonate L)/**26** (not in 4 male neonates, 5 neonates, 2 male fetuses, 9 female neonates, 2 female fetuses)	0/**7** (not in female fetus, 2 male fetuses, female 2.3 yrs, female 2.4 yrs, female 24 yrs, female 29 yrs)
Musculus sacrococcygeus anterior (from sacrum to coccyx)	0/**20** (not in 8 male neonates, 6 neonates, male fetus, 3 female neonates, male 6 yrs, female fetus)	0/**17** (not in 2 male neonates, 5 neonates, male fetus, 8 female neonates, female fetus)	1 (female 2.3 yrs L)/**5** (not in female 2.4 yrs, female 24 yrs, male fetus, female fetus)
Pectoralis minimus (from first rib to coracoid process of scapula)	0/**24** (not in 10 male neonates, 6 neonates, male fetus, 5 female neonates, male 6 yrs, female fetus)	9 (male neonate B, female neonate L, female neonate R, female neonate L, male neonate B, female neonate B, male neonate R, female fetus L, female neonate L)/**26** (not in 2 male neonates, 5 neonates, 2 male fetuses, 7 female neonates, female fetus)	0/**7** (not in female fetus, 2 male fetuses, female 2.3 yrs, female 2.4 yrs, female 24 yrs, female 29 yrs)

Table 3 contd....

Table 3 contd.

Muscle anomalies	Trisomy 13 (presence versus absence, shown in bold)	Trisomy 18 (presence versus absence, shown in bold)	Trisomy 21 (presence versus absence, shown in bold)
Supernumerary			
"Pectoralis abdominalis"	0/24 (not in 10 male neonates, 6 neonates, male fetus, 5 female neonates, male 6 yrs, female fetus)	0/26 (not in 5 male neonates, 5 neonates, 2 male fetuses, 12 female neonates, 2 female fetuses)	1 (male fetus R)/7 (not in female fetus, male fetus, female 2.3 yrs, female 2.4 yrs, female 24 yrs, female 29 yrs)
Panniculus carnosus (cutaneous muscle sheet arising from pectoral muscle mass and covering various trunk regions)	0/24 (not in 10 male neonates, 6 neonates, male fetus, 5 female neonates, male 6 yrs, female fetus)	1 (female fetus B)/26 (not in 5 male neonates, 5 neonates, 2 male fetuses, 12 female neonates, female fetus)	1 (male fetus B)/7 (not in female fetus, male fetus, female 2.3 yrs, female 2.4 yrs, female 24 yrs, female 29 yrs)
Sternochondroscapularis (manubrium and 1st rib to broad aponeurosis on coracoid process and head of humerus)	1 (female neonate R)/24 (not in 10 male neonates, 6 neonates, male fetus, 4 female neonates, male 6 yrs, female fetus)	0/26 (not in 5 male neonates, 5 neonates, 2 male fetuses, 12 female neonates, 2 female fetuses)	0/7 (not in female fetus, 2 male fetuses, female 2.3 yrs, female 2.4 yrs, female 24 yrs, female 29 yrs)
Accessory brachialis	0/24 (not in 10 male neonates, 6 neonates, male fetus, 5 female neonates, male 6 yrs, female fetus)	1 (female fetus L)/26 (not in 5 male neonates, 5 neonates, 2 male fetuses, 12 female neonates, female fetus)	0/7 (not in female fetus, 2 male fetuses, female 2.3 yrs, female 2.4 yrs, female 24 yrs, female 29 yrs)
Coracobrachialis profundus/brevis (extra deep belly coracobrachialis)	2 (female neonate L, female neonate L)/24 (not in 10 male neonates, 6 neonates, male fetus, 3 female neonates, male 6 yrs, female fetus)	0/26 (not in 5 male neonates, 5 neonates, 2 male fetuses, 12 female neonates, 2 female fetuses)	0/7 (not in female fetus, 2 male fetuses, female 2.3 yrs, female 2.4 yrs, female 24 yrs, female 29 yrs)
Coracobrachialis longus/superficialis (coracoid process to distal humerus)	0/24 (not in 10 male neonates, 6 neonates, male fetus, 5 female neonates, male 6 yrs, female fetus)	2 (female neonate L, female neonate B)/26 (not in 5 male neonates, 5 neonates, 2 male fetuses, 10 female neonates, 2 female fetuses)	0/7 (not in female fetus, 2 male fetuses, female 2.3 yrs, female 2.4 yrs, female 24 yrs, female 29 yrs)
Chondroepitrochlearis (tendon of latissimus dorsi to medial epicondyle of humerus)	5 (male neonate B, female neonate L, female neonate L, neonate, neonate)/24 (not in 9 male neonates, 4 neonates, male fetus, 3 female neonates, male 6 yrs, female fetus)	0/26 (not in 5 male neonates, 5 neonates, 2 male fetuses, 12 female neonates, 2 female fetuses)	0/7 (not in female fetus, 2 male fetuses, female 2.3 yrs, female 2.4 yrs, female 24 yrs, female 29 yrs)
Tensor semivaginae scapulohumeralis (manubrium to humeral head/scapula)	1 (female neonate R)/24 (not in 10 male neonates, 6 neonates, male fetus, 4 female neonates, male 6 yrs, female fetus)	10 (male neonate B, female neonate B, female neonate B, female neonate B, female neonate B, female neonate B, male fetus B, male neonate L, female fetus B, female neonate L)/26 (not in 3 male neonates, 5 neonates, male fetus, 6 female neonates, female fetus)	0/7 (not in female fetus, 2 male fetuses, female 2.3 yrs, female 2.4 yrs, female 24 yrs, female 29 yrs)

Table 3 contd....

Table 3 contd.

Muscle anomalies	Trisomy 13 (presence versus absence, shown in bold)	Trisomy 18 (presence versus absence, shown in bold)	Trisomy 21 (presence versus absence, shown in bold)
Supernumerary			
Slip between coracoid process and 5th rib at midaxillary line	**0/24** (not in 10 male neonates, 6 neonates, male fetus, 5 female neonates, male 6 yrs, female fetus)	1 (female neonate L)/**26** (not in 5 male neonates, 5 neonates, 2 male fetuses, 11 female neonates, 2 female fetuses)	**0/7** (not in female fetus, 2 male fetuses, female 2.3 yrs, female 2.4 yrs, female 24 yrs, female 29 yrs)
Epitrochleoanconeus (from the back of the medial condyle of the humerus to the medial side of the olecranon process)	**0/24** (not in 10 male neonates, 6 neonates, male fetus, 5 female neonates, male 6 yrs, female fetus)	2 (male neonate L, male fetus L)/**26** (not in 4 male neonates, 5 neonates, male fetus, 12 female neonates, 2 female fetuses)	**0/7** (not in female fetus, 2 male fetuses, female 2.3 yrs, female 2.4 yrs, female 24 yrs, female 29 yrs)
Radiocarpus (radius to flexor retinaculum)	5 (female neonate R, male neonate R, female neonate B, male neonate B, male neonate L)/**24** (not in 7 male neonates, 6 neonates, male fetus, 3 female neonates, male 6 yrs, female fetus)	5 (female neonate R, female neonate B, male neonate R, female fetus L, female neonate R)/**26** (not in 4 male neonates, 5 neonates, 2 male fetuses, 9 female neonates, female fetus)	**0/7** (not in female fetus, 2 male fetuses, female 2.3 yrs, female 2.4 yrs, female 24 yrs, female 29 yrs)
Supernumerary forearm flexors (innervated by musculocutaneous nerve)	**0/24** (not in 10 male neonates, 6 neonates, male fetus, 5 female neonates, male 6 yrs, female fetus)	**0/26** (not in 5 male neonates, 5 neonates, 2 male fetuses, 12 female neonates, 2 female fetuses)	3 (female 24 yrs B, male fetus L, female 2.3 yrs)/**7** (not in female fetus, male fetus, female 2.4 yrs, female 29 yrs)
Accessory radiocarpus (originating from underside of flexor digitorum superficialis belly and inserting onto radiocarpus belly)	2 (male neonate R, male neonate L)/**24** (not in 8 male neonates, 6 neonates, male fetus, 5 female neonates, male 6 yrs, female fetus)	**0/26** (not in 5 male neonates, 5 neonates, 2 male fetuses, 12 female neonates, 2 female fetuses)	**0/7** (not in female fetus, 2 male fetuses, female 2.3 yrs, female 2.4 yrs, female 24 yrs, female 29 yrs)
Extensor medii digitii	**0/24** (not in 10 male neonates, 6 neonates, male fetus, 5 female neonates, male 6 yrs, female fetus)	5 (female neonate B (et quarti R), female neonate B, male fetus L, male neonate B, female fetus L, female neonate B, female neonate B)/**26** (not in 4 male neonates, 5 neonates, male fetus, 8 female neonates, female fetus)	**0/7** (not in female fetus, 2 male fetuses, female 2.3 yrs, female 2.4 yrs, female 24 yrs, female 29 yrs)
Flexor carpi ulnaris accessorius (originating from middle of ulna and inserting with digiti minimi onto 5th digit)	**0/24** (not in 10 male neonates, 6 neonates, male fetus, 5 female neonates, male 6 yrs, female fetus)	9 (male neonate B, female neonate B, female neonate B, female neonate L, female neonate B, female neonate R, female neonate L, male neonate L, female neonate L)/**26** (not in 3 male neonates, 5 neonates, 2 male fetuses, 5 female neonates, 2 female fetuses)	**0/7** (not in female fetus, 2 male fetuses, female 2.3 yrs, female 2.4 yrs, female 24 yrs, female 29 yrs)

Table 3 contd....

Table 3 contd.

Muscle anomalies	Trisomy 13 (presence versus absence, shown in bold)	Trisomy 18 (presence versus absence, shown in bold)	Trisomy 21 (presence versus absence, shown in bold)
Supernumerary			
Extensor carpi radialis accessorius (superficial to extensor carpi radialis)	0/24 (not in 10 male neonates, 6 neonates, male fetus, 5 female neonates, male 6 yrs, female fetus)	**5** (female neonate B, male neonate B, female neonate B, male neonate L, female neonate L)/**26** (not in 3 male neonates, 5 neonates, 2 male fetuses, 9 female neonates, 2 female fetuses)	0/7 (not in female fetus, 2 male fetuses, female 2.3 yrs, female 2.4 yrs, female 24 yrs, female 29 yrs)
Flexor carpi radialis brevis (originating from flexor carpi radialis belly, inserting onto trapezium)	0/24 (not in 10 male neonates, 6 neonates, male fetus, 5 female neonates, male 6 yrs, female fetus)	**1** (female neonate B)/**26** (not in 5 male neonates, 5 neonates, 2 male fetuses, 11 female neonates, 2 female fetuses)	0/7 (not in female fetus, 2 male fetuses, female 2.3 yrs, female 2.4 yrs, female 24 yrs, female 29 yrs)
Accessory flexor digitorum profundus	**1** (male neonate L)/**24** (not in 9 male neonates, 6 neonates, male fetus, 5 female neonates, male 6 yrs, female fetus)	0/**26** (not in 5 male neonates, 5 neonates, 2 male fetuses, 12 female neonates, 2 female fetuses)	0/7 (not in female fetus, 2 male fetuses, female 2.3 yrs, female 2.4 yrs, female 24 yrs, female 29 yrs)
Accessory flexor pollicis longus (medial epicondyle of humerus to flexor pollicis brevis)	**1** (female fetus B)/**24** (not in 10 male neonates, 6 neonates, male fetus, 5 female neonates, male 6 yrs)	0/**26** (not in 5 male neonates, 5 neonates, 2 male fetuses, 12 female neonates, 2 female fetuses)	0/7 (not in female fetus, 2 male fetuses, female 2.3 yrs, female 2.4 yrs, female 24 yrs, female 29 yrs)
"Deep head of Cruveilhier" (which is however often a synonym of the interosseous volaris primus of Henle)	**2** (female neonate B, female neonate R)/**24** (not in 10 male neonates, 6 neonates, male fetus, 3 female neonates, male 6 yrs, female fetus)	**3** (female neonate B, female neonate R, male fetus R)/**26** (not in 5 male neonates, 5 neonates, male fetus, 10 female neonates, 2 female fetuses)	**2** (female 29 yrs B, male fetus B)/**7** (not in female fetus, male fetus, female 2.3 yrs, female 2.4 yrs, female 24 yrs)
Interosseous volaris primus of Henle (metacarpal of 1st digit to the same metacarpal and/or proximal phalange of 1st digit)	**5** (male neonate B, male neonate B, female neonate B, female neonate B, female neonate L)/**24** (not in 8 male neonates, 6 neonates, male fetus, 2 female neonates, male 6 yrs, female fetus)	**10** (female neonate B, female neonate B, female neonate R, female neonate B, female neonate B, male neonate B, male neonate B, male fetus B, male neonate B, female fetus B)/**26** (not in 2 male neonates, 5 neonates, male fetus, 7 female neonates, female fetus)	**2** (female 29 yrs B, male fetus B)/**7** (not in female fetus, male fetus, female 2.3 yrs, female 2.4 yrs, female 24 yrs)
Contrahens to digit II	0/24 (not in 10 male neonates, 6 neonates, male fetus, 5 female neonates, male 6 yrs, female fetus)	**4** (female neonate L, female neonate B, female neonate B, female neonate L)/**26** (not in 5 male neonates, 5 neonates, 2 male fetuses, 8 female neonates, 2 female fetuses)	0/7 (not in female fetus, 2 male fetuses, female 2.3 yrs, female 2.4 yrs, female 24 yrs, female 29 yrs)
Contrahens to digit IV	0/24 (not in 10 male neonates, 6 neonates, male fetus, 5 female neonates, male 6 yrs, female fetus)	**1** (female neonate L)/**26** (not in 5 male neonates, 5 neonates, 2 male fetuses, 11 female neonates, 2 female fetuses)	0/7 (not in female fetus, 2 male fetuses, female 2.3 yrs, female 2.4 yrs, female 24 yrs, female 29 yrs)

Table 3 contd....

Table 3 contd.

Muscle anomalies	Trisomy 13 (presence versus absence, shown in bold)	Trisomy 18 (presence versus absence, shown in bold)	Trisomy 21 (presence versus absence, shown in bold)
Supernumerary			
Contrahens to digit V	**0/24** (not in 10 male neonates, 6 neonates, male fetus, 5 female neonates, male 6 yrs, female fetus)	**2** (female neonate R, female fetus L)/**26** (not in 5 male neonates, 5 neonates, 2 male fetuses, 11 female neonates, female fetus)	**0/7** (not in female fetus, 2 male fetuses, female 2.3 yrs, female 2.4 yrs, female 24 yrs, female 29 yrs)
"Extra lumbricalis" between digits I and II	**0/24** (not in 10 male neonates, 6 neonates, male fetus, 5 female neonates, male 6 yrs, female fetus)	**5** (female neonate B, female neonate L, male fetus R, male neonate L, female fetus B)/**26** (not in 4 male neonates, 5 neonates, male fetus, 10 female neonates, female fetus)	**0/7** (not in female fetus, 2 male fetuses, female 2.3 yrs, female 2.4 yrs, female 24 yrs, female 29 yrs)
Tenuissimus (sacrum/coccyx and sacrotuberous ligament to iliotibial tract)	**0/20** (not in 8 male neonates, 6 neonates, male fetus, 3 female neonates, male 6 yrs, female fetus)	**2** (female neonate L, female neonate R)/**17** (not in 2 male neonates, 5 neonates, male fetus, 6 female neonates, female fetus)	**0/5** (not in female fetus, male fetus, female 2.3 yrs, female 2.4 yrs, female 24 yrs)
Belly between semitendinosus and long head of biceps femoris	**0/20** (not in 8 male neonates, 6 neonates, male fetus, 3 female neonates, male 6 yrs, female fetus)	**3** (male neonate R, female neonate R, female neonate R)/**17** (not in male neonate, 5 neonates, male fetus, 6 female neonates, female fetus)	**0/5** (not in female fetus, male fetus, female 2.3 yrs, female 2.4 yrs, female 24 yrs)
Belly between sacrotuberous lig. and semitendinosus	**0/20** (not in 8 male neonates, 6 neonates, male fetus, 3 female neonates, male 6 yrs, female fetus)	**2** (female neonate L, female neonate R)/**17** (not in 2 male neonates, 5 neonates, male fetus, 6 female neonates, female fetus)	**0/5** (not in female fetus, male fetus, female 2.3 yrs, female 2.4 yrs, female 24 yrs)
Accessory vastus lateralis	**1** (male fetus B)/**20** (not in 8 male neonates, 6 neonates, 3 female neonates, male 6 yrs, female fetus)	**0/17** (not in 2 male neonates, 5 neonates, male fetus, 8 female neonates, female fetus)	**0/5** (not in female fetus, male fetus, female 2.3 yrs, female 2.4 yrs, female 24 yrs)
Fibulocalcaneus medialis	**2** (male fetus B, male 6 yrs B)/**20** (not in 8 male neonates, 6 neonates, 3 female neonates, female fetus)	**0/17** (not in 2 male neonates, 5 neonates, male fetus, 8 female neonates, female fetus)	**1** (female 2.4 yrs B)/**5** (not in female 2.3 yrs, female 24 yrs, male fetus, female fetus)
Peroneocalcaneus medialis (tibia to underside of calcaneus)	**0/20** (not in 8 male neonates, 6 neonates, male fetus, 3 female neonates, male 6 yrs, female fetus)	**3** (male neonate B, female neonate B, female neonate B)/**17** (not in male neonate, 5 neonates, male fetus, 6 female neonates, female fetus)	**0/5** (not in female fetus, male fetus, female 2.3 yrs, female 2.4 yrs, female 24 yrs)
Extensor primi internodii hallucis (proximal portion of tibia to proximal phalanx of hallux)	**0/20** (not in 8 male neonates, 6 neonates, male fetus, 3 female neonates, male 6 yrs, female fetus)	**1** (female neonate B)/**17** (not in 2 male neonates, 5 neonates, male fetus, 7 female neonates, female fetus)	**0/5** (not in female fetus, male fetus, female 2.3 yrs, female 2.4 yrs, female 24 yrs)

Table 3 contd....

Table 3 contd.

Muscle anomalies	Trisomy 13 (presence versus absence, shown in bold)	Trisomy 18 (presence versus absence, shown in bold)	Trisomy 21 (presence versus absence, shown in bold)
Supernumerary			
Peroneus quartus (distal fibula to calcaneus)	**3** (male neonate B, male neonate B, male 6 yrs B)/**20** (not in 6 male neonates, 6 neonates, male fetus, 3 female neonates, female fetus)	**5** (male neonate L, female neonate B, female neonate L, female neonate L, female neonate R)/**17** (not in male neonate, 5 neonates, male fetus, 4 female neonates, female fetus)	**0/5** (not in female fetus, male fetus, female 2.3 yrs, female 2.4 yrs, female 24 yrs)
Peroneus digiti quinti (distal fibula to 5th metatarsal/phalanges)	**2** (neonate, neonate)/**20** (not in 8 male neonates, 4 neonates, male fetus, 3 female neonates, male 6 yrs, female fetus)	**3** (female neonate B, neonate, neonate)/**17** (not in 2 male neonates, 3 neonates, male fetus, 7 female neonates, female fetus)	**0/5** (not in female fetus, male fetus, female 2.3 yrs, female 2.4 yrs, female 24 yrs)
Ulnaris digiti quinti	**1** (female neonate B)/**20** (not in 8 male neonates, 6 neonates, male fetus, 2 female neonates, male 6 yrs, female fetus)	**0/17** (not in 2 male neonates, 5 neonates, male fetus, 8 female neonates, female fetus)	**0/5** (not in female fetus, male fetus, female 2.3 yrs, female 2.4 yrs, female 24 yrs)
Absent			
Auricularis anterior	**0/20** (present in 8 male neonates, 6 neonates, male fetus, 3 female neonates, male 6 yrs, female fetus)	**1** (female neonate B)/**17** (present in 2 male neonates, 5 neonates, male fetus, 7 female neonates, female fetus)	**0/5** (present in female fetus, male fetus, female 2.3 yrs, female 2.4 yrs, female 24 yrs)
Risorius	**0/20** (present in 8 male neonates, 6 neonates, male fetus, 3 female neonates, male 6 yrs, female fetus)	**1** (female neonate B)/**17** (present in 2 male neonates, 5 neonates, male fetus, 7 female neonates, female fetus)	**0/5** (present in female fetus, male fetus, female 2.3 yrs, female 2.4 yrs, female 24 yrs)
Pectoralis minor	**3** (female neonate L, male neonate R, female neonate L)/**24** (present in 9 male neonates, 6 neonates, male fetus, 3 female neonates, male 6 yrs, female fetus)	**1** (male neonate L)/**26** (present in 4 male neonates, 5 neonates, 2 male fetuses, 12 female neonates, 2 female fetuses)	**0/7** (present in female fetus, 2 male fetuses, female 2.3 yrs, female 2.4 yrs, female 24 yrs, female 29 yrs)
Stylohyoideus	**5** (female neonate B, female fetus L, neonate, neonate, male neonate R)/**20** (present in 7 male neonates, 4 neonates, male fetus, 2 female neonates, male 6 yrs)	**7** (male fetus L, male neonate L, female neonate B, female neonate B, female neonate L, male neonate L, neonate)/**17** (present in 4 neonates, 5 female neonates, female fetus)	**0/5** (present in female fetus, male fetus, female 2.3 yrs, female 2.4 yrs, female 24 yrs)
Serratus posterior inferior	**4** (female neonate B, female fetus B, male fetus L, male neonate B)/**24** (present in 9 male neonates, 6 neonates, 4 female neonates, male 6 yrs)	**1** (female neonate B)/**26** (present in 5 male neonates, 5 neonates, 2 male fetuses, 11 female neonates, 2 female fetuses)	**0/7** (present in female fetus, 2 male fetuses, female 2.3 yrs, female 2.4 yrs, female 24 yrs, female 29 yrs)
Serratus posterior superior	**1** (male neonate R)/**24** (present in 9 male neonates, 6 neonates, male fetus, 5 female neonates, male 6 yrs, female fetus)	**0/26** (present in 5 male neonates, 5 neonates, 2 male fetuses, 12 female neonates, 2 female fetuses)	**0/7** (present in female fetus, 2 male fetuses, female 2.3 yrs, female 2.4 yrs, female 24 yrs, female 29 yrs)

Table 3 contd....

Table 3 contd.

Muscle anomalies	Trisomy 13 (presence versus absence, shown in bold)	Trisomy 18 (presence versus absence, shown in bold)	Trisomy 21 (presence versus absence, shown in bold)
Absent			
Subclavius	10 (male neonate L, male neonate B, female neonate B, female neonate R, female neonate B, male 6 yrs R, male neonate L, male neonate L, female fetus L, male fetus L)/20 (not in 4 male neonates, 6 neonates)	11 (female neonate L, female neonate R, female neonate L, female neonate L, female neonate B, female neonate B, female neonate B, male fetus R, male neonate L, female fetus B)/17 (present in male neonate, 5 neonates)	0/5 (present in female fetus, male fetus, female 2.3 yrs, female 2.4 yrs, female 24 yrs)
Abductor pollicis longus	0/24 (present in 10 male neonates, 6 neonates, male fetus, 5 female neonates, male 6 yrs, female fetus)	1 (female neonate R)/26 (present in 5 male neonates, 5 neonates, 2 male fetuses, 11 female neonates, 2 female fetuses)	0/7 (present in female fetus, 2 male fetuses, female 2.3 yrs, female 2.4 yrs, female 24 yrs, female 29 yrs)
Abductor pollicis brevis	1 (female neonate L)/24 (present in 10 male neonates, 6 neonates, male fetus, 4 female neonates, male 6 yrs, female fetus)	9 (male neonate R, female neonate R, female neonate B, female neonate L, female neonate B, female neonate B, male neonate B, female fetus L, female neonate B)/26 (present in 3 male neonates, 5 neonates, 2 male fetuses, 6 female neonates, female fetus)	0/7 (present in female fetus, 2 male fetuses, female 2.3 yrs, female 2.4 yrs, female 24 yrs, female 29 yrs)
Flexor pollicis brevis	1 (female neonate L)/24 (present in 10 male neonates, 6 neonates, male fetus, 4 female neonates, male 6 yrs, female fetus)	8 (male neonate R, female neonate R, female neonate R, female neonate L, female neonate B, female neonate B, male neonate B, female neonate B)/26 (present in 3 male neonates, 5 neonates, 2 male fetuses, 6 female neonates, 2 female fetuses)	0/7 (present in female fetus, 2 male fetuses, female 2.3 yrs, female 2.4 yrs, female 24 yrs, female 29 yrs)
Flexor pollicis longus	0/24 (present in 10 male neonates, 6 neonates, male fetus, 5 female neonates, male 6 yrs, female fetus)	1 (female neonate R)/26 (present in 5 male neonates, 5 neonates, 2 male fetuses, 11 female neonates, 2 female fetuses)	0/7 (present in female fetus, 2 male fetuses, female 2.3 yrs, female 2.4 yrs, female 24 yrs, female 29 yrs)
Extensor pollicis brevis	0/24 (present in 10 male neonates, 6 neonates, male fetus, 5 female neonates, male 6 yrs, female fetus)	4 (female neonate R, male fetus B, female fetus L, female fetus L)/26 (present in 5 male neonates, 5 neonates, male fetus, 11 female neonates)	0/7 (present in female fetus, 2 male fetuses, female 2.3 yrs, female 2.4 yrs, female 24 yrs, female 29 yrs)
Extensor indicis	1 (female neonate R)/24 (present in 10 male neonates, 6 neonates, male fetus, 4 female neonates, male 6 yrs, female fetus)	0/26 (present in 5 male neonates, 5 neonates, 2 male fetuses, 12 female neonates, 2 female fetuses)	0/7 (present in female fetus, 2 male fetuses, female 2.3 yrs, female 2.4 yrs, female 24 yrs, female 29 yrs)

Table 3 contd....

Table 3 contd.

Muscle anomalies	Trisomy 13 (presence versus absence, shown in bold)	Trisomy 18 (presence versus absence, shown in bold)	Trisomy 21 (presence versus absence, shown in bold)
Absent			
Tendon of extensor digitorum superficialis to 5th digit	0/24 (present in 10 male neonates, 6 neonates, male fetus, 5 female neonates, male 6 yrs, female fetus)	1 (female fetus B)/26 (present in 5 male neonates, 5 neonates, 2 male fetuses, 12 female neonates, female fetus)	0/7 (present in female fetus, 2 male fetuses, female 2.3 yrs, female 2.4 yrs, female 24 yrs, female 29 yrs)
Short head, biceps brachii	2 (female neonate L, male fetus L)/24 (present in 10 male neonates, 6 neonates, 4 female neonates, male 6 yrs, female fetus)	0/26 (present in 5 male neonates, 5 neonates, 2 male fetuses, 12 female neonates, 2 female fetuses)	0/7 (present in female fetus, 2 male fetuses, female 2.3 yrs, female 2.4 yrs, female 24 yrs, female 29 yrs)
Long head, biceps brachii	0/24 (present in 10 male neonates, 6 neonates, male fetus, 5 female neonates, male 6 yrs, female fetus)	1 (female neonate B)/26 (present in 5 male neonates, 5 neonates, 2 male fetuses, 11 female neonates, 2 female fetuses)	1 (female 24 yrs B)/7 (present in female fetus, 2 male fetuses, female 2.3 yrs, female 2.4 yrs, female 29 yrs)
Palmaris longus	20 (male neonate B, male neonate B, male neonate B, female neonate B, female neonate B, female neonate B, male neonate B, male neonate B, female neonate B, male neonate B, male neonate B, female fetus B, male fetus B, neonate, neonate, neonate, female neonate B, male neonate B, male neonate B, male neonate B)/24 (present in 3 neonates, male 6 yrs)	23 (female neonate B, female neonate B, female neonate B, female neonate B, female neonate B, female neonate B, male neonate B, male neonate B, male fetus B, female fetus L, female neonate B, female neonate B, female neonate B, male neonate L, male neonate B, male fetus B, male neonate B, female fetus L, female neonate B, female neonate B, neonate, neonate)/26 (present in 3 neonates)	6 (female 24 yrs B, female 2.4 yrs B, female 2.3 yrs B, male fetus B, female fetus B, female 29 yrs)/7 (present in male fetus)
Palmaris brevis	20 (male neonate B, male neonate B, male neonate B, female neonate B, male neonate B, male neonate B, female neonate B, female neonate B, female neonate B, male neonate B, male neonate B, female fetus B, male fetus B, neonate, neonate, neonate, female neonate B, male neonate B, male neonate B, male neonate B)/24 (present in 3 neonates, male 6 yrs)	15 (female neonate B, female neonate L, female neonate B, female neonate B, female neonate B, female neonate B, female neonate B, female neonate R, male neonate B, male neonate B, male fetus B, male neonate B, female fetus B, female neonate B, neonate, neonate)/26 (present in 2 male neonates, 3 neonates, male fetus, 4 female neonates, female fetus)	1 (male fetus L)/7 (present in female fetus, male fetus, female 2.3 yrs, female 2.4 yrs, female 24 yrs, female 29 yrs)
Flexor digitorum superficialis, radial head	4 (female neonate R, female neonate R, male neonate B, male neonate B)/24 (present in 8 male neonates, 6 neonates, male fetus, 3 female neonates, male 6 yrs, female fetus)	2 (male fetus R, female fetus L)/26 (present in 5 male neonates, 5 neonates, male fetus, 12 female neonates, female fetus)	0/7 (present in female fetus, 2 male fetuses, female 2.3 yrs, female 2.4 yrs, female 24 yrs, female 29 yrs)

Table 3 contd....

Table 3 contd.

Muscle anomalies	Trisomy 13 (presence versus absence, shown in bold)	Trisomy 18 (presence versus absence, shown in bold)	Trisomy 21 (presence versus absence, shown in bold)
Absent			
Tendon of flexor digitorum superficialis to 5th digit	7 (male neonate B, female neonate R, female neonate L, male fetus L, female neonate B, male neonate B, male 6 yrs R, hypoplastic on L)/**24** (present in 8 male neonates, 6 neonates, 2 female neonates, female fetus)	9 (female neonate L, male neonate B, female neonate B, female neonate B, female neonate B, female fetus L, female neonate B, female neonate R, male neonate B)/**26** (present in 3 male neonates, 5 neonates, 2 male fetuses, 6 female neonates, female fetus)	3 (female fetus L, female 24 yrs R, male fetus B)/**7** (present in male fetus, female 2.3 yrs, female 2.4 yrs, female 29 yrs)
Tendon of flexor digitorum profundus to 5th digit	1 (male neonate)/**24** (present in 9 male neonates, 6 neonates, male fetus, 5 female neonates, male 6 yrs, female fetus)	1 (female neonate)/**26** (present in 5 male neonates, 5 neonates, 2 male fetuses, 11 female neonates, 2 female fetuses)	**0/7** (present in female fetus, 2 male fetuses, female 2.3 yrs, female 2.4 yrs, female 24 yrs, female 29 yrs)
Extensor digiti minimi	1 (female neonate B)/**24** (present in 10 male neonates, 6 neonates, male fetus, 4 female neonates, male 6 yrs, female fetus)	2 (female neonate B, female neonate B)/**26** (present in 5 male neonates, 5 neonates, 2 male fetuses, 10 female neonates, 2 female fetuses)	**0/7** (present in female fetus, 2 male fetuses, female 2.3 yrs, female 2.4 yrs, female 24 yrs, female 29 yrs)
Extensor carpi radialis brevis	**0/24** (present in 10 male neonates, 6 neonates, male fetus, 5 female neonates, male 6 yrs, female fetus)	2 (female neonate B, female neonate B)/**26** (present in 5 male neonates, 5 neonates, 2 male fetuses, 10 female neonates, 2 female fetuses)	**0/7** (present in female fetus, 2 male fetuses, female 2.3 yrs, female 2.4 yrs, female 24 yrs, female 29 yrs)
Flexor carpi radialis	**0/24** (present in 10 male neonates, 6 neonates, male fetus, 5 female neonates, male 6 yrs, female fetus)	3 (female neonate L, female neonate B, female neonate B)/**26** (present in 5 male neonates, 5 neonates, 2 male fetuses, 9 female neonates, 2 female fetuses)	**0/7** (present in female fetus, 2 male fetuses, female 2.3 yrs, female 2.4 yrs, female 24 yrs, female 29 yrs)
Extensor indicis	3 (female fetus B, female neonate L, male neonate R)/**24** (present in 9 male neonates, 6 neonates, 4 female neonates, male 6 yrs, female fetus)	**0/26** (present in 5 male neonates, 5 neonates, 2 male fetuses, 12 female neonates, 2 female fetuses)	**0/7** (present in female fetus, 2 male fetuses, female 2.3 yrs, female 2.4 yrs, female 24 yrs, female 29 yrs)
Superficial head, flexor pollicis brevis	1 (male neonate R)/**24** (present in 9 male neonates, 6 neonates, male fetus, 5 female neonates, male 6 yrs, female fetus)	**0/26** (present in 5 male neonates, 5 neonates, 2 male fetuses, 12 female neonates, 2 female fetuses)	**0/7** (present in female fetus, 2 male fetuses, female 2.3 yrs, female 2.4 yrs, female 24 yrs, female 29 yrs)
Opponens pollicis	1 (male fetus R)/**24** (present in 10 male neonates, 6 neonates, 5 female neonates, male 6 yrs, female fetus)	6 (female neonate B, female neonate B, female neonate L, female neonate B, female neonate B, female neonate B)/**26** (present in 5 male neonates, 5 neonates, 2 male fetuses, 6 female neonates, 2 female fetuses)	**0/7** (present in female fetus, 2 male fetuses, female 2.3 yrs, female 2.4 yrs, female 24 yrs, female 29 yrs)

Table 3 contd....

Table 3 contd.

Muscle anomalies	Trisomy 13 (presence versus absence, shown in bold)	Trisomy 18 (presence versus absence, shown in bold)	Trisomy 21 (presence versus absence, shown in bold)
Absent			
Adductor pollicis, transverse head	**1** (male neonate R)/**24** (present in 9 male neonates, 6 neonates, male fetus, 5 female neonates, male 6 yrs, female fetus)	**2** (female neonate B (whole muscle), female neonate L)/**26** (present in 5 male neonates, 5 neonates, 2 male fetuses, 10 female neonates, 2 female fetuses)	**0/7** (present in female fetus, 2 male fetuses, female 2.3 yrs, female 2.4 yrs, female 24 yrs, female 29 yrs)
Lumbricalis 1	**3** (female neonate B, male neonate R, male fetus B)/**24** (present in 9 male neonates, 6 neonates, 4 female neonates, male 6 yrs, female fetus)	**4** (female neonate L, female neonate L, male fetus L, female fetus R)/**26** (present in 5 male neonates, 5 neonates, male fetus, 10 female neonates, female fetus)	**0/7** (present in female fetus, 2 male fetuses, female 2.3 yrs, female 2.4 yrs, female 24 yrs, female 29 yrs)
Lumbricalis 2	**0/24** (present in 10 male neonates, 6 neonates, male fetus, 5 female neonates, male 6 yrs, female fetus)	**2** (female neonate L, female neonate L)/**26** (present in 5 male neonates, 5 neonates, 2 male fetuses, 10 female neonates, 2 female fetuses)	**0/7** (present in female fetus, 2 male fetuses, female 2.3 yrs, female 2.4 yrs, female 24 yrs, female 29 yrs)
Lumbricalis 3	**0/24** (present in 10 male neonates, 6 neonates, male fetus, 5 female neonates, male 6 yrs, female fetus)	**2** (female neonate L, female neonate L)/**26** (present in 5 male neonates, 5 neonates, 2 male fetuses, 10 female neonates, 2 female fetuses)	**0/7** (present in female fetus, 2 male fetuses, female 2.3 yrs, female 2.4 yrs, female 24 yrs, female 29 yrs)
Lumbricalis 4	**1** (male neonate R)/**24** (present in 9 male neonates, 6 neonates, male fetus, 5 female neonates, male 6 yrs, female fetus)	**2** (female neonate L, female neonate L)/**26** (present in 5 male neonates, 5 neonates, 2 male fetuses, 10 female neonates, 2 female fetuses)	**0/7** (present in female fetus, 2 male fetuses, female 2.3 yrs, female 2.4 yrs, female 24 yrs, female 29 yrs)
First palmar and dorsal interossei	**0/24** (present in 10 male neonates, 6 neonates, male fetus, 5 female neonates, male 6 yrs, female fetus)	**2** (female neonate L, female neonate R)/**26** (present in 5 male neonates, 5 neonates, 2 male fetuses, 10 female neonates, 2 female fetuses)	**0/7** (present in female fetus, 2 male fetuses, female 2.3 yrs, female 2.4 yrs, female 24 yrs, female 29 yrs)
Pronator quadratus	**0/24** (present in 10 male neonates, 6 neonates, male fetus, 5 female neonates, male 6 yrs, female fetus)	**1** (female neonate L)/**26** (present in 5 male neonates, 5 neonates, 2 male fetuses, 11 female neonates, 2 female fetuses)	**0/7** (present in female fetus, 2 male fetuses, female 2.3 yrs, female 2.4 yrs, female 24 yrs, female 29 yrs)
Gemellus superior	**1** (male neonate R)/**24** (present in 9 male neonates, 6 neonates, male fetus, 5 female neonates, male 6 yrs, female fetus)	**3** (male neonate B, female neonate L, female neonate B)/**26** (present in 4 male neonates, 5 neonates, 2 male fetuses, 10 female neonates, 2 female fetuses)	**0/7** (present in female fetus, 2 male fetuses, female 2.3 yrs, female 2.4 yrs, female 24 yrs, female 29 yrs)
Psoas minor	**0/20** (present in 8 male neonates, 6 neonates, male fetus, 3 female neonates, male 6 yrs, female fetus)	**9** (male neonate B, male neonate B, female neonate B, female neonate B, female neonate B, female neonate B, female neonate B, female	**3** (female 2.4 yrs B, female 2.3 yrs L, male fetus B)/**5** (present in female fetus, female 24 yrs)

Table 3 contd....

Table 3 contd.

Muscle anomalies	Trisomy 13 (presence versus absence, shown in bold)	Trisomy 18 (presence versus absence, shown in bold)	Trisomy 21 (presence versus absence, shown in bold)
Absent			
		neonate B, female neonate B)/**17** (present in 5 neonates, male fetus, 6 female neonates, female fetus)	
Piriformis	**0/20** (present in 8 male neonates, 6 neonates, male fetus, 3 female neonates, male 6 yrs, female fetus)	**1** (female neonate L)/**17** (present in 2 male neonates, 5 neonates, male fetus, 7 female neonates, female fetus)	**0/5** (present in female fetus, male fetus, female 2.3 yrs, female 2.4 yrs, female 24 yrs)
Plantaris	**8** (female neonate B, male neonate L, male neonate B, male fetus B, female neonate B, male neonate B, male neonate B, male neonate B)/**20** (not in 3 male neonates, 6 neonates, female neonate, male 6 yrs, female fetus)	**1** (female fetus B)/**17** (present in 2 male neonates, 5 neonates, male fetus, 8 female neonates)	**0/5** (present in female fetus, male fetus, female 2.3 yrs, female 2.4 yrs, female 24 yrs)
Fibularis tertius	**12** (female neonate B, male neonate B, male neonate B, female fetus B, male fetus B, neonate, neonate, neonate, female neonate B, male neonate B, male neonate B, male neonate B)/**20** (not in 3 male neonates, 3 neonates, female neonate, male 6 yrs)	**2** (neonate, neonate)/**17** (present in 2 male neonates, 3 neonates, male fetus, 8 female neonates, female fetus)	**0/5** (present in female fetus, male fetus, female 2.3 yrs, female 2.4 yrs, female 24 yrs)
Tibialis anterior	**1** (female neonate B)/**20** (present in 8 male neonates, 6 neonates, male fetus, 2 female neonates, male 6 yrs, female fetus)	**0/17** (present in 2 male neonates, 5 neonates, male fetus, 8 female neonates, female fetus)	**0/5** (present in female fetus, male fetus, female 2.3 yrs, female 2.4 yrs, female 24 yrs)
Transverse head, adductor hallucis	**1** (male neonate L, male fetus R)/**20** (present in 7 male neonates, 6 neonates, 3 female neonates, male 6 yrs, female fetus)	**0/17** (present in 2 male neonates, 5 neonates, male fetus, 8 female neonates, female fetus)	**0/5** (present in female fetus, male fetus, female 2.3 yrs, female 2.4 yrs, female 24 yrs)
Quadratus plantae	**1** (lateral head absent in male 6 yrs B)/**20** (present in 8 male neonates, 6 neonates, male fetus, 3 female neonates, female fetus)	**0/17** (present in 2 male neonates, 5 neonates, male fetus, 8 female neonates, female fetus)	**0/5** (present in female fetus, male fetus, female 2.3 yrs, female 2.4 yrs, female 24 yrs)
Tendon of flexor digitorum brevis to 5th digit	**5** (male fetus B, male neonate L, male neonate B, female fetus B, female neonate B)/**20** (present in 6 male neonates, 6 neonates, 2 female neonates, male 6 yrs)	**7** (female neonate L, female neonate L, female neonate R, male neonate B, female neonate B, female neonate B, female neonate B)/**17** (present in male neonate, 5 neonates, 2 female neonates, male fetus, female fetus)	**2** (female 24 yrs R, male fetus B)/**5** (present in female 2.4 yrs, female 2.3 yrs, female fetus)

Table 3 contd....

Table 3 contd.

Muscle anomalies	Trisomy 13 (presence versus absence, shown in bold)	Trisomy 18 (presence versus absence, shown in bold)	Trisomy 21 (presence versus absence, shown in bold)
Variant Muscles			
Poor differentiation of mid-facial muscles	**0/20** (normal in 8 male neonates, 6 neonates, male fetus, 3 female neonates, male 6 yrs, female fetus)	**0/17** (normal in 2 male neonates, 5 neonates, male fetus, 8 female neonates, female fetus)	**5** (female 24 yrs B, female 2.4 yrs B, female 2.3 yrs B, male fetus B, female fetus B)/**5**
Platysma myoides	**2** (left and right sides separated, no cervical part in neonate, neonate)/**20** (normal in 8 male neonates, 4 neonates, male fetus, 3 female neonates, male 6 yrs, female fetus)	**1** (originating above clavicles in female neonate B)/**17** (normal in 2 male neonates, 5 neonates, male fetus, 7 female neonates, female fetus)	**0/5** (normal in female fetus, male fetus, female 2.3 yrs, female 2.4 yrs, female 24 yrs)
Digastricus posterior	**5** (attaches onto hyoid without communicating with anterior belly in male fetus B, merged w/stylohyoid in male 6 yrs B, doubled and inserted onto hyoid in male neonate B, doubled and left are inserted onto hyoid, superomedial of posterior bellies continuous with lateral of the anterior bellies, the right inferolateral of the two posterior bellies inserted into infrahyoid muscles in female fetus B, doubled in male neonate, from styloid process and one from mastoid process on left and split bellies on right blending with stylopharyngeus B)/**20** (normal in 6 male neonates, 6 neonates, 3 female neonates)	**4** (extra muscle sips to styloid process in male neonate L, female neonate B, female neonate B, female neonate L)/**17** (normal in male neonate, 5 neonates, male fetus, 5 female neonates, female fetus)	**0/5** (normal in female fetus, male fetus, female 2.3 yrs, female 2.4 yrs, female 24 yrs)
Pterygoideus lateralis	**0/20** (normal in 8 male neonates, 6 neonates, male fetus, 3 female neonates, male 6 yrs, female fetus)	**1** (attached to deep aspect of TMJ and adjacent temporal bone in female neonate B)/**17** (normal in 2 male neonates, 5 neonates, male fetus, 7 female neonates, female fetus)	**0/5** (normal in female fetus, male fetus, female 2.3 yrs, female 2.4 yrs, female 24 yrs)

Table 3 contd....

Table 3 contd.

Muscle anomalies	Trisomy 13 (presence versus absence, shown in bold)	Trisomy 18 (presence versus absence, shown in bold)	Trisomy 21 (presence versus absence, shown in bold)
Variant Muscles			
Digastricus anterior	**4** (two accessory slips to mandible and one to masseter on R while two accessory slips were on L in female neonate, extra fasiculus joining mylohyoid in male neonate L, medial anterior belly arose separately from hyoid bone, both right anterior bellies insert on mandible with no connections to mylohyoid in female fetus B, doubled bilaterally in male neonate B)/**20** (normal in 6 male neonates, 6 neonates, male fetus, 2 female neonates, male 6 yrs)	**5** (both anterior bellies fused across midline in male fetus B, large in female neonate B, doubled bilaterally in female neonate B, extra slip from hyoid to mandible in neonate B, doubled in female neonate R)/**17** (normal in 2 male neonates, 4 neonates, 5 female neonates, female fetus)	**0/5** (normal in female fetus, male fetus, female 2.3 yrs, female 2.4 yrs, female 24 yrs)
Mylohyoideus	**2** (lacked median raphe and undirected fibers B and lacked posterior portion on R in female neonate, deficient anteriorly in male neonate, B)/**20** (normal in 7 male neonates, 6 neonates, male fetus, 2 female neonates, male 6 yrs, female fetus)	**2** (extra muscle slip bilaterally in female neonate B, lacks median raphe in female neonate B)/**17** (normal in 2 male neonates, 5 neonates, male fetus, 6 female neonates, female fetus)	**0/5** (normal in female fetus, male fetus, female 2.3 yrs, female 2.4 yrs, female 24 yrs)
Stylohyoideus	**2** (did not bifurcate around digastric and blended w/ it in both male neonate B, and male fetus B)/**20** (normal in 7 male neonates, 6 neonates, 3 female neonates, male 6 yrs, female fetus)	**5** (male neonate R, female neonate B, female neonate R, male neonate R, female neonate B)/**17** (normal in 5 neonates, male fetus, 5 female neonates, female fetus)	**0/5** (normal in female fetus, male fetus, female 2.3 yrs, female 2.4 yrs, female 24 yrs)
Stylopharyngeus	**1** (doubled to insert just lateral to pharyngeal raphe in male neonate B)/**20** (normal in 7 male neonates, 6 neonates, male fetus, 3 female neonates, male 6 yrs, female fetus)	**0/17** (normal in 2 male neonates, 5 neonates, male fetus, 8 female neonates, female fetus)	**0/5** (normal in female fetus, male fetus, female 2.3 yrs, female 2.4 yrs, female 24 yrs)
Geniohyoideus	**0/20** (normal in 8 male neonates, 6 neonates, male fetus, 3 female neonates, male 6 yrs, female fetus)	**1** (doubled bilaterally in female neonate B)/**17** (normal in 2 male neonates, 5 neonates, male fetus, 7 female neonates, female fetus)	**0/5** (normal in female fetus, male fetus, female 2.3 yrs, female 2.4 yrs, female 24 yrs)

Table 3 contd....

Table 3 contd.

Muscle anomalies	Trisomy 13 (presence versus absence, shown in bold)	Trisomy 18 (presence versus absence, shown in bold)	Trisomy 21 (presence versus absence, shown in bold)
Variant Muscles			
Hyoglossus	**1** (comprised of two distinct sheets separated by a gap in male fetus R)/**20** (normal in 8 male neonates, 6 neonates, 3 female neonates, male 6 yrs, female fetus)	0/**17** (normal in 2 male neonates, 5 neonates, male fetus, 8 female neonates, female fetus)	0/**5** (normal in female fetus, male fetus, female 2.3 yrs, female 2.4 yrs, female 24 yrs)
Vertebrocostal trigones	**2** (female neonate R, male neonate B)/**20** (normal in 7 male neonates, 6 neonates, male fetus, 2 female neonates, male 6 yrs, female fetus)	0/**17** (normal in 2 male neonates, 5 neonates, male fetus, 8 female neonates, female fetus)	0/**5** (normal in female fetus, male fetus, female 2.3 yrs, female 2.4 yrs, female 24 yrs)
Deltopectoral complex (fused deltoid and pectoralis major)	0/**24** (not in 10 male neonates, 6 neonates, male fetus, 5 female neonates, male 6 yrs, female fetus)	**21** (female neonate B, female neonate B, female neonate B, female neonate B, female neonate B, female neonate B, male neonate B, male neonate B, female neonate B, female neonate B, female neonate B, female neonate B, male neonate B, male neonate B, male fetus B, female fetus B female neonate B, female neonate L, neonate, neonate, neonate B)/**26** (not in male neonate, 2 neonates, male fetus, female fetus)	0/**7** (not in female fetus, 2 male fetuses, female 2.3 yrs, female 2.4 yrs, female 24 yrs, female 29 yrs)
Deltoideus (only)	**1** (divided into clavicular, acromial and spinous portions in female neonate L)/**24** (normal in 10 male neonates, 6 neonates, male fetus, 4 female neonates, male 6 yrs, female fetus)	0/**26** (normal in 5 male neonates, 5 neonates, 2 male fetuses, 12 female neonates, 2 female fetuses)	0/**7** (normal in female fetus, 2 male fetuses, female 2.3 yrs, female 2.4 yrs, female 24 yrs, female 29 yrs)
Slip from pterygoideus lateralis to pterygoideus medialis	**1** (From superior fasiculus of lateral pterygoid to near the insertion point of medial pterygoid in male neonate B)/**20** (normal in 7 male neonates, 6 neonates, male fetus, 3 female neonates, male 6 yrs, female fetus)	0/**17** (normal in 2 male neonates, 5 neonates, male fetus, 8 female neonates, female fetus)	0/**5** (normal in female fetus, male fetus, female 2.3 yrs, female 2.4 yrs, female 24 yrs)
Sternohyoideus	**1** (doubled in male 6yrs L)/**20** (normal in 8 male neonates, 6 neonates, male fetus, 3 female neonates, female fetus)	**4** (accessory slip in female neonate L, doubled bilaterally in female neonate and female neonate B, doubled in female neonate L)/**17** (normal in 2 male neonates, 5 neonates, male fetus, 4 female neonates, female fetus)	0/**5** (normal in female fetus, male fetus, female 2.3 yrs, female 2.4 yrs, female 24 yrs)

Table 3 contd....

Table 3 contd.

Muscle anomalies	Trisomy 13 (presence versus absence, shown in bold)	Trisomy 18 (presence versus absence, shown in bold)	Trisomy 21 (presence versus absence, shown in bold)
Variant Muscles			
Sternothyroideus	1 (left extension joined extra belly of posterior digastric in male neonate)/ 20 (normal in 7 male neonates, 6 neonates, male fetus, 3 female neonates, male 6 yrs, female fetus)	0/17 (normal in 2 male neonates, 5 neonates, male fetus, 8 female neonates, female fetus)	0/5 (normal in female fetus, male fetus, female 2.3 yrs, female 2.4 yrs, female 24 yrs)
Omohyoideus	7 (Intermediate tendon absent in neonate, fused with sternohyoideus in neonate, extra superior belly in male neonate, fused with sternohyoideus in neonate B, neonate B, neonate B, hypoplastic in male fetus R)/24 (normal in 9 male neonates, neonate, 5 female neonates, male 6 yrs, female fetus)	11 (lacks intermediate tendon in female neonate B, superior belly in female neonate, intermediate bellies absent in neonate B, neonate B, poorly developed intermediate tendon in male neonate, female neonate, female neonate, female neonate, male neonate, female neonate, female neonate, no middle tendon and no clavicular insertion in female neonate B)/26 (normal in 3 male neonates, 3 neonates, 2 male fetuses, 4 female neonates, 2 female fetuses)	0/7 (normal in female fetus, 2 male fetuses, female 2.3 yrs, female 2.4 yrs, female 24 yrs, female 29 yrs)
Sternocleidomastoideus	1 (three bellies, lateral belly inserts on manubrium, middle onto sternoclavicular junction, lateral onto clavicle in male neonate B)/20 (normal in 7 male neonates, 6 neonates, male fetus, 3 female neonates, male 6 yrs, female fetus)	2 (divided into sterno and cleido portions in female neonate L, separated into sterno and cliedomastoid portions in neonate B)/17 (normal in 2 male neonates, 4 neonates, male fetus, 7 female neonates, female fetus)	0/5 (normal in female fetus, male fetus, female 2.3 yrs, female 2.4 yrs, female 24 yrs)
Obliquus capitis superior	1 (extra belly in female fetus L)/20 (normal in 8 male neonates, 6 neonates, male fetus, 3 female neonates, male 6 yrs)	0/17 (normal in 2 male neonates, 5 neonates, male fetus, 8 female neonates, female fetus)	0/5 (normal in female fetus, male fetus, female 2.3 yrs, female 2.4 yrs, female 24 yrs)
Obliquus capitis inferior	1 (extra belly on left extra slip on right joining semispinalis cervicis in male neonate B)/20 (normal in 7 male neonates, 6 neonates, male fetus, 3 female neonates, male 6 yrs, female fetus)	0/17 (normal in 2 male neonates, 5 neonates, male fetus, 8 female neonates, female fetus)	0/5 (normal in female fetus, male fetus, female 2.3 yrs, female 2.4 yrs, female 24 yrs)

Table 3 contd....

Table 3 contd.

Muscle anomalies	Trisomy 13 (presence versus absence, shown in bold)	Trisomy 18 (presence versus absence, shown in bold)	Trisomy 21 (presence versus absence, shown in bold)
Variant Muscles			
Scalenus posterior	**0/24** (normal in 10 male neonates, 6 neonates, male fetus, 5 female neonates, male 6 yrs, female fetus)	**1** (origin blended with levator scapulae in female neonate B)/**26** (normal in 5 male neonates, 5 neonates, 2 male fetuses, 11 female neonates, 2 female fetuses)	**0/7** (normal in female fetus, 2 male fetuses, female 2.3 yrs, female 2.4 yrs, female 24 yrs, female 29 yrs)
Rhomboideus (major and minor)	**0/24** (normal in 10 male neonates, 6 neonates, male fetus, 5 female neonates, male 6 yrs, female fetus)	**2** (major and minor fused in female neonate B, female neonate B)/**26** (normal in 5 male neonates, 5 neonates, 2 male fetuses, 10 female neonates, 2 female fetuses)	**0/7** (normal in female fetus, 2 male fetuses, female 2.3 yrs, female 2.4 yrs, female 24 yrs, female 29 yrs)
Undersurface of diaphragm and right crus	**1** (slip from right crus to psoas major, slip present on central tendon, slip from thoracic surface of left portion of central tendon to pericardium just below left inferior pulmonary vein in male neonate R)/**20** (normal in 7 male neonates, 6 neonates, male fetus, 3 female neonates, male 6 yrs, female fetus)	**0/17** (normal in 2 male neonates, 5 neonates, male fetus, 8 female neonates, female fetus)	**0/5** (normal in female fetus, male fetus, female 2.3 yrs, female 2.4 yrs, female 24 yrs)
Subscapularis	**0/24** (normal in 10 male neonates, 6 neonates, male fetus, 5 female neonates, male 6 yrs, female fetus)	**1** (insertion split plane of muscle in female neonate B)/**26** (normal in 5 male neonates, 5 neonates, 2 male fetuses, 11 female neonates, 2 female fetuses)	**0/7** (normal in female fetus, 2 male fetuses, female 2.3 yrs, female 2.4 yrs, female 24 yrs, female 29 yrs)
Pectoralis minor	**0/24** (normal in 10 male neonates, 6 neonates, male fetus, 5 female neonates, male 6 yrs, female fetus)	**10** (slip from 5th rib in male neonate B, female neonate R, female neonate B, origin split in male neonate L, male neonate L, inserts on humerus in male neonate R, male fetus R, female fetus L, lacks origin from 2nd rib and is doubled in female neonate L, partial insertion into short head of biceps brachii in female neonate B)/**26** (normal in male neonate, 5 neonates, male fetus, 8 female neonates, female fetus)	**0/7** (normal in female fetus, 2 male fetuses, female 2.3 yrs, female 2.4 yrs, female 24 yrs, female 29 yrs)

Table 3 contd....

Table 3 contd.

Muscle anomalies	Trisomy 13 (presence versus absence, shown in bold)	Trisomy 18 (presence versus absence, shown in bold)	Trisomy 21 (presence versus absence, shown in bold)
Variant Muscles			
Pectoralis major	5 (Abdominal head absent in male neonate B, male neonate B, female neonate B, female neonate B, divided into clavicular, sternal, and costal portions in female neonate L)/24 (normal in 8 male neonates, 6 neonates, male fetus, 2 female neonates, male 6 yrs, female fetus)	7 (Clavicular head absent in female neonate B, male neonate L, female neonate B, clavicular portion absent, attached to deltoid and biceps brachii on left and small portion inserted into biceps brachii on right in female neonate B, inserted on biceps brachii in neonate, neonate, neonate)/26 (normal in 4 male neonates, 2 neonates, 2 male fetuses, 9 female neonates, 2 female fetuses)	0/7 (normal in female fetus, 2 male fetuses, female 2.3 yrs, female 2.4 yrs, female 24 yrs, female 29 yrs)
Coracobrachialis	1 (split into brevis and medius in female neonate L)/24 (normal in 10 male neonates, 6 neonates, male fetus, 4 female neonates, male 6 yrs, female fetus)	3 (represented by small bellies in addition to short head of biceps in female neonate B, and female neonate B, fused with short head of biceps in female neonate R)/26 (normal in 5 male neonates, 5 neonates, 2 male fetuses, 9 female neonates, 2 female fetuses)	0/7 (normal in female fetus, 2 male fetuses, female 2.3 yrs, female 2.4 yrs, female 24 yrs, female 29 yrs)
Biceps brachii	12 (Accessory heads in female neonate B, Extra tendon to coracoacromial lig R and no attachment to coracoid process R and fleshy slip inserted on to pronator teres B. while on left the short head arose from coracoacromial lig and third head arose from joint capsule L in male 6 yrs, accessory heads in male neonate B, accessory heads and muscle slips between biceps and brachialis in neonate, neonate, neonate, slip to brachialis in female neonate R, accessory head L and short head diminutive and short head humeral origin in female neonate B, accessory head R and short head humeral origin in male neonate L)/24 (normal in 8 male neonates, male fetus, 2 female neonates, female fetus)	15 (short head fused with coracobrachialis and divided in female neonate B, female neonate R, female neonate B, long head absent in female neonate B, Accessory heads in female neonate B, female neonate B, male neonate B, male neonate L, accessory heads B and coracobrachialis medius slip to long head of biceps in female neonate B, accessory heads B and portion fused with deltopectoral in female neonate B, deltopectoral complex inserts onto biceps in female neonate B, female fetus L and accessory heads L in that female fetus, long head humeral origin in female neonate R, female neonate L, female neonate B)/26 (normal in 3 male neonates, 5 neonates, 2 male fetuses, female fetus)	1 (accessory heads in female 29 yrs L)/7 (normal in female 2.3 yrs, female fetus, 2 male fetuses, female 2.4 yrs, female 24 yrs)

Table 3 contd....

Table 3 contd.

Muscle anomalies	Trisomy 13 (presence versus absence, shown in bold)	Trisomy 18 (presence versus absence, shown in bold)	Trisomy 21 (presence versus absence, shown in bold)
Variant Muscles			
Triceps brachii	**2** (more proximal origin in female neonate B, female neonate R)/**24** (normal in 10 male neonates, 6 neonates, male fetus, 3 female neonates, male 6 yrs, female fetus)	**18** (medial head extends beyond radial groove of humerus to same level in female neonate B, female neonate B, female neonate B, female neonate B, female neonate B, female neonate B, male neonate B, male neonate B, high humeral origin in male fetus B, more proximal origin in female neonate B, female neonate B, female neonate B, female neonate B, female neonate B, male neonate B, male neonate B, male fetus B, and slip from deltoid in male neonate B, Female fetus B)/**26** (normal in 5 neonates, female neonate, 2 female fetuses)	**0/7** (normal in female fetus, 2 male fetuses, female 2.3 yrs, female 2.4 yrs, female 24 yrs, female 29 yrs)
Brachialis	**2** (two accessory slips joining biceps brachii in female neonate R, female neonate R)/**24** (normal in 10 male neonates, 6 neonates, male fetus, 3 female neonates, male 6 yrs, female fetus)	**3** (high origin in male fetus L, male neonate L, divided into two portions in female neonate R)/**26** (normal in 4 male neonates, 5 neonates, male fetus, 11 female neonates, 2 female fetuses)	**0/7** (normal in female fetus, 2 male fetuses, female 2.3 yrs, female 2.4 yrs, female 24 yrs, female 29 yrs)
Brachioradialis	**0/24** (normal in 10 male neonates, 6 neonates, male fetus, 5 female neonates, male 6 yrs, female fetus)	**3** (attaches more distally on radialis in female neonate L, diminutive in male fetus B, male neonate L)/**26** (normal in 4 male neonates, 5 neonates, male fetus, 11 female neonates, 2 female fetuses)	**0/7** (normal in female fetus, 2 male fetuses, female 2.3 yrs, female 2.4 yrs, female 24 yrs, female 29 yrs)
Accessory tendon/slip from flexor digitorum superficialis to flexor digitorum profundus	**3** (female neonate L, male neonate L, female neonate B)/**24** (not in 9 male neonates, 6 neonates, male fetus, 3 female neonates, male 6 yrs, female fetus)	**7** (female neonate B, female neonate R, female neonate L, female neonate B, male neonate B, male fetus L, male neonate R)/**26** (not in 3 male neonates, 5 neonates, male fetus, 8 female neonates, 2 female fetuses)	**2** (female fetus B, female 24 yrs B)/**7** (not in 2 male fetuses, female 2.3 yrs, female 2.4 yrs, female 29 yrs)
Accessory tendon from flexor digitorum superficialis to lumbricalis 1	**0/24** (not in 10 male neonates, 6 neonates, male fetus, 5 female neonates, male 6 yrs, female fetus)	**0/26** (not in 5 male neonates, 5 neonates, 2 male fetuses, 12 female neonates, 2 female fetuses)	**1** (female 2.3 yrs B)/**7** (not in female fetus, 2 male fetuses, female 2.4 yrs, female 24 yrs, female 29 yrs)

Table 3 contd....

Table 3 contd.

Muscle anomalies	Trisomy 13 (presence versus absence, shown in bold)	Trisomy 18 (presence versus absence, shown in bold)	Trisomy 21 (presence versus absence, shown in bold)
Variant Muscles			
Accessory tendon from flexor pollicis longus to flexor digitorum profundus	0/**24** (not in 10 male neonates, 6 neonates, male fetus, 5 female neonates, male 6 yrs, female fetus)	0/**26** (not in 5 male neonates, 5 neonates, 2 male fetuses, 12 female neonates, 2 female fetuses)	2 (female fetus R, male fetus B)/**7** (not in male fetus, female 2.3 yrs, female 2.4 yrs, female 24 yrs, female 29 yrs)
Accessory slip from flexor digitorum superficialis to flexor pollicis longus	3 (female neonate L, male neonate L, male neonate R)/**24** (not in 8 male neonates, 6 neonates, male fetus, 4 female neonates, male 6 yrs, female fetus)	8 (female neonate L, female neonate B, female neonate B, female neonate L, male neonate R, male fetus L, male neonate L, female fetus B)/**26** (not in 3 male neonates, 5 neonates, male fetus, 8 female neonates, female fetus)	1 (male fetus B)/**7** (not in female fetus, male fetus, female 2.3 yrs, female 2.4 yrs, female 24 yrs, female 29 yrs)
Accessory slip from flexor digitorum superficialis to radiocarpus	1 (female neonate R)/**24** (not in 10 male neonates, 6 neonates, male fetus, 4 female neonates, male 6 yrs, female fetus)	1 (female neonate R)/**26** (not in 5 male neonates, 5 neonates, 2 male fetuses, 11 female neonates, 2 female fetuses)	0/**7** (not in female fetus, 2 male fetuses, female 2.3 yrs, female 2.4 yrs, female 24 yrs, female 29 yrs)
Flexor pollicis longus	2 (insertion also on metacarpal 1 in male neonate R, female neonate L)/**24** (normal in 9 male neonates, 6 neonates, male fetus, 4 female neonates, male 6 yrs, female fetus)	13 (tendon split into three slips in female neonate B, attached to base of metacarpal 2 in female neonate R, sent aponeurotic expansions to whole length of lateral border of metacarpal 1 and to lateral side of base of proximal phalanx of thumb in female neonate B, tendon to index finger in male fetus B and female fetus B, insertion also on metacarpal 1 in female neonate B, female neonate B, male fetus L, male neonate R, tendons to digits 1 and 2 in female neonate R, female neonate B, female fetus L)/**26** (normal in 4 male neonates, 5 neonates, 5 female neonates)	1 (tendons to digits 1 and 2 in male fetus B)/**7** (normal in female 2.3 yrs, female fetus, male fetus, female 2.4 yrs, female 24 yrs, female 29 yrs)
Flexor pollicis brevis	2 (fused with opponens pollicis in female neonate L, male neonate B)/**24** (normal in 9 male neonates, 6 neonates, male fetus, 4 female neonates, male 6 yrs, female fetus)	10 (diminutive in female neonate L, female neonate B, female neonate R, male fetus L, fused with opponens pollicis in female neonate B, male neonate B, male neonate L, female fetus L, ulnar head in female neonate B, male neonate B)/**26** (normal in 2 male neonates, 5 neonates, male fetus, 7 female neonates, female fetus)	0/**7** (normal in female fetus, 2 male fetuses, female 2.3 yrs, female 2.4 yrs, female 24 yrs, female 29 yrs)

Table 3 contd....

Table 3 contd.

Muscle anomalies	Trisomy 13 (presence versus absence, shown in bold)	Trisomy 18 (presence versus absence, shown in bold)	Trisomy 21 (presence versus absence, shown in bold)
Variant Muscles			
Extensor carpi radialis (longus and brevis)	**3** (longus and brevis fused in male neonate B, female fetus B, longus inserted onto second metacarpal distal to its base while brevis inserted onto third metacarpal distal it's base in female neonate B)/**24** (normal in 9 male neonates, 6 neonates, male fetus, 4 female neonates, male 6 yrs)	**10** (longus attached to styloid process in female neonate B, tendon doubled in male neonate B, longus and brevis fused in male neonate B, female neonate B, longus doubled in female neonate B (and inserts onto abductor pollicis brevis B), longus and brevis doubled in male neonate B, female fetus B(and brevis inserts distally onto metacarpal 3 L), longus and brevis fused in female neonate B, female neonate B, male fetus L)/**26** (normal in 2 male neonates, 5 neonates, male fetus, 7 female neonates, female fetus)	**0/7** (normal in female fetus, 2 male fetuses, female 2.3 yrs, female 2.4 yrs, female 24 yrs, female 29 yrs)
Extensor carpi ulnaris	**1** (inserted onto medial aspect of 5th metacarpal distal it's base in female neonate B)/**24** (normal in 10 male neonates, 6 neonates, male fetus, 4 female neonates, male 6 yrs, female fetus)	**0/26** (normal in 5 male neonates, 5 neonates, 2 male fetuses, 12 female neonates, 2 female fetuses)	**0/7** (normal in female fetus, 2 male fetuses, female 2.3 yrs, female 2.4 yrs, female 24 yrs, female 29 yrs)
Flexor carpi ulnaris	**0/24** (normal in 10 male neonates, 6 neonates, male fetus, 5 female neonates, male 6 yrs, female fetus)	**1** (accessory head in male fetus B)/**26** (normal in 5 male neonates, 5 neonates, male fetus, 12 female neonates, 2 female fetuses)	**0/7** (normal in female fetus, 2 male fetuses, female 2.3 yrs, female 2.4 yrs, female 24 yrs, female 29 yrs)
Flexor carpi radialis	**0/24** (normal in 10 male neonates, 6 neonates, male fetus, 5 female neonates, male 6 yrs, female fetus)	**2** (tendon attached to proximal border of flexor retinaculum in male neonate R, attached to base of metacarpal I in female neonate R)/**26** (normal in 4 male neonates, 5 neonates, 2 male fetuses, 11 female neonates, 2 female fetuses)	**0/7** (normal in female fetus, 2 male fetuses, female 2.3 yrs, female 2.4 yrs, female 24 yrs, female 29 yrs)
Abductor pollicis longus	**4** (male neonate B, male neonate B, female neonate B, female neonate R)/**24** (normal in 8 male neonates, 6 neonates, male fetus, 3 female neonates, male 6 yrs, female fetus)	**9** (doubled in female neonate B, attaches to flexor retinaculum in female neonate B and female neonate L, female neonate B, male neonate L, female fetus B, male neonate L, male neonate L, female neonate B)/**26** (normal in 2 male neonates, 5 neonates, 2 male fetuses, 7 female neonates, female fetus)	**0/7** (normal in female fetus, 2 male fetuses, female 2.3 yrs, female 2.4 yrs, female 24 yrs, female 29 yrs)

Table 3 contd....

Table 3 contd.

Muscle anomalies	Trisomy 13 (presence versus absence, shown in bold)	Trisomy 18 (presence versus absence, shown in bold)	Trisomy 21 (presence versus absence, shown in bold)
Variant Muscles			
Abductor pollicis brevis	**0/24** (normal in 10 male neonates, 6 neonates, male fetus, 5 female neonates, male 6 yrs, female fetus)	**5** (diminutive in female neonate L, female neonate B, male neonate B, female neonate R, male fetus L)/**26** (normal in 4 male neonates, 5 neonates, male fetus, 9 female neonates, 2 female fetuses)	**0/7** (normal in female fetus, 2 male fetuses, female 2.3 yrs, female 2.4 yrs, female 24 yrs, female 29 yrs)
Opponens pollicis	**1** (diminutive in female neonate L)/**24**(normal in 10 male neonates, 6 neonates, male fetus, 4 female neonates, male 6 yrs, female fetus)	**3** (diminutive in female neonate R, male neonate R, female fetus B)/**26**(normal in 4 male neonates, 5 neonates, 2 male fetuses, 11 female neonates, female fetus)	**0/7** (normal in female fetus, 2 male fetuses, female 2.3 yrs, female 2.4 yrs, female 24 yrs, female 29 yrs)
Extensor indicis	**7** (originated from near the distal radioulnar joint in male neonate B and male fetus B, diminutive in male neonate B, female neonate L, originated from distal end of ulna on R while on the L it gives off 2 tendons one to middle and one to distal phalanx on radial aspect and diminutive in female neonate B, originated from dorsal aspect of medial carpal bones in male 6 yrs B)/**24** (normal in 8 male neonates, 6 neonates, 3 female neonates, female fetus)	**12** (doubled in female neonate B, female neonate L, female neonate R, male neonate B, female neonate B, tendon doubled in male fetus R, doubled in female neonate B, female neonate B, male neonate B, male neonate B, male neonate B, female fetus B)/**26** (normal in male neonate, 5 neonates, male fetus, 6 female neonates, female fetus)	**2** (doubled in female 29 yrs L, male fetus L)/ **7** (normal in female fetus, male fetus, female 2.4 yrs, female 24 yrs, female 2.3 yrs)
Abductor digiti minimi	**1** (inserts onto digit VI in male neonate L)/**24** (normal in 9 male neonates, 6 neonates, male fetus, 5 female neonates, male 6 yrs, female fetus)	**0/26** (normal in 5 male neonates, 5 neonates, 2 male fetuses, 12 female neonates, 2 female fetuses)	**0/7** (normal in female fetus, 2 male fetuses, female 2.3 yrs, female 2.4 yrs, female 24 yrs, female 29 yrs)
Lumbricalis 1	**0/24** (normal in 10 male neonates, 6 neonates, male fetus, 5 female neonates, male 6 yrs, female fetus)	**3** (arose from ulnar side of tendon of fl. Digitorum profundus to index finger and fused with 2nd lumbricalis in female neonate B, arose from ulnar side of fl. pollicis longus over 1st metacarpal joint in female neonate L, and female neonate L)/**26** (normal in 5 male neonates, 5 neonates, 2 male fetuses, 9 female neonates, 2 female fetuses)	**0/7** (normal in female fetus, 2 male fetuses, female 2.3 yrs, female 2.4 yrs, female 24 yrs, female 29 yrs)

Table 3 contd....

Table 3 contd.

Muscle anomalies	Trisomy 13 (presence versus absence, shown in bold)	Trisomy 18 (presence versus absence, shown in bold)	Trisomy 21 (presence versus absence, shown in bold)
Variant Muscles			
Lumbricalis 4	**0**/24 (normal in 10 male neonates, 6 neonates, male fetus, 5 female neonates, male 6 yrs, female fetus)	**1** (tendon entered synovial sheath of the little finger deep to the tendon of fl. Digitorum profundus and extended to an attachment at the base of 2nd phalanx within sheath in female neonate B)/**26** (normal in 5 male neonates, 5 neonates, 2 male fetuses, 11 female neonates, 2 female fetuses)	**0**/7 (normal in female fetus, 2 male fetuses, female 2.3 yrs, female 2.4 yrs, female 24 yrs, female 29 yrs)
Palmar interosseous 4 (origin from opponens digiti minimi)	**1** (male neonate B)/**24** (normal in 9 male neonates, 6 neonates, male fetus, 5 female neonates, male 6 yrs, female fetus)	**0**/**26** (normal in 5 male neonates, 5 neonates, 2 male fetuses, 12 female neonates, 2 female fetuses)	**0**/7 (normal in female fetus, 2 male fetuses, female 2.3 yrs, female 2.4 yrs, female 24 yrs, female 29 yrs)
Extensor digitorum	**1** (extra slips/tendons in male neonate R)/**24** (normal in 10 male neonates, 6 neonates, male fetus, 5 female neonates, male 6 yrs, female fetus)	**9** (extensor hood displaced to either radial or ulnar side in female neonate B, female neonate B, female neonate B, female neonate B, female neonate B, male neonate B, male neonate B, extra fascicle going to digiti quinti in female neonate B)/**26** (normal in 3 male neonates, 5 neonates, 2 male fetuses, 5 female neonates, 2 female fetuses)	**0**/7 (normal in female fetus, 2 male fetuses, female 2.3 yrs, female 2.4 yrs, female 24 yrs, female 29 yrs)
Extensor pollicis (longus and brevis)	**2** (longus hypoplastic in male 6 yrs B, longus and brevis fusion in female neonate R)/**24** (normal in 10 male neonates, 6 neonates, male fetus, 4 female neonates, female fetus)	**13** (doubling of tendon in female neonate B, brevis split into three parts in female neonate B, brevis split into two parts in male neonate L, attaches to styloid process of radius in female neonate R, attaches further distally than usual to distal phalanx in female neonate R, female neonate B, longus and brevis fused in female neonate B, longus sends tendon to index finger in female fetus L, longus inserts onto proximal phalanx and brevis is diminutive in female fetus L, longus and brevis	**1** (et indicis longus in female 29 yrs L)/**7** (normal in female fetus, 2 male fetuses, female 2.3 yrs, female 2.4 yrs, female 24 yrs)

Table 3 contd....

Table 3 contd.

Muscle anomalies	Trisomy 13 (presence versus absence, shown in bold)	Trisomy 18 (presence versus absence, shown in bold)	Trisomy 21 (presence versus absence, shown in bold)
Variant Muscles			
		fused in female neonate L, male neonate B, male fetus L, doubled tendon inserts on distal phalanx in female neonate B)/**26** (normal in 3 male neonates, 5 neonates, male fetus, 4 female neonates)	
Palmaris longus	**0**/**24** (normal in 10 male neonates, 6 neonates, male fetus, 5 female neonates, male 6 yrs, female fetus)	**2** (diminutive in male neonate R, digastric and insertion on metacarpal 5 in female fetus)/**26** (normal in 4 male neonates, 5 neonates, 2 male fetuses, 12 female neonates, female fetus)	**2** (belly distally located in female 29 yrs R, diminutive in male fetus R)/**7** (normal in female fetus, male fetus, female 2.3 yrs, female 2.4 yrs, female 24 yrs)
Palmaris brevis	**1** (ran obliquely from pisiform to flexor retinaculum and laid deep to ulnar artery and nerve in male neonate R)/**24** (normal in 9 male neonates, 6 neonates, male fetus, 5 female neonates, male 6 yrs, female fetus)	**0**/**26** (normal in 5 male neonates, 5 neonates, 2 male fetuses, 12 female neonates, 2 female fetuses)	**0**/**7** (normal in female fetus, 2 male fetuses, female 2.3 yrs, female 2.4 yrs, female 24 yrs, female 29 yrs)
Lumbricales	**5** (4th lumbricalis origin also from flexor digitorum superficialis V tendon in male neonate R, 4th lumbricalis inserts on digits IV and V in male neonate B, female neonate L, 3rd lumbricalis inserts onto digits III and IV in female neonate L, female neonate L)/**24** (normal in 8 male neonates, 6 neonates, male fetus, 2 female neonates, male 6 yrs, female fetus)	**9** (3rd lumbricalis inserts onto digits III and IV in female neonate B, female neonate R, male neonate R (also 3rd lumbricalis sends slip to digit V R, and 4th lumbricalis inserts onto flexor digitorum superficialis V tendon L), originates off flexor digitorum superficialis II, IV and V in female neonate R (also in the next female neonate B, female neonate B, male fetus R), originates off of flexor pollicis longus in female neonate R (also 4th lumbricalis inserts on digits IV and V L), male neonate B, male fetus L, female fetus L, abnormal in female neonate B)/**26** (normal in 3 male neonates, 5 neonates, male fetus, 6 female neonates, female fetus)	**1** (double between digits 3–4 and 4–5 on right in hand double between digits 4–5 in foot L female 2.3 yrs B)/**7** (normal in female fetus, 2 male fetuses, female 2.4 yrs, female 24 yrs, female 29 yrs)

Table 3 contd....

Table 3 contd.

Muscle anomalies	Trisomy 13 (presence versus absence, shown in bold)	Trisomy 18 (presence versus absence, shown in bold)	Trisomy 21 (presence versus absence, shown in bold)
Variant Muscles			
Flexor digiti minimi	1 (extra slips/tendons in female neonate R)/24 (normal in 10 male neonates, 6 neonates, male fetus, 4 female neonates, male 6 yrs, female fetus)	0/26 (normal in 5 male neonates, 5 neonates, 2 male fetuses, 12 female neonates, 2 female fetuses)	0/7 (normal in female fetus, 2 male fetuses, female 2.3 yrs, female 2.4 yrs, female 24 yrs, female 29 yrs)
Flexor hallucis brevis	1 (extra slips/tendons in male neonate L)/20 (normal in 7 male neonates, 6 neonates, male fetus, 3 female neonates, male 6 yrs, female fetus)	0/17 (normal in 2 male neonates, 5 neonates, male fetus, 8 female neonates, female fetus)	0/5 (normal in female fetus, male fetus, female 2.3 yrs, female 2.4 yrs, female 24 yrs)
Flexor hallucis longus	0/20 (normal in 8 male neonates, 6 neonates, male fetus, 3 female neonates, male 6 yrs, female fetus)	2 (tendon to navicular bone and slip to fl. Digitorum longus in male fetus L, tendon to calcaneus and slip to fl. Digitorum longus in female fetus R)/17 (normal in 2 male neonates, 5 neonates, 8 female neonates)	0/5 (normal in female fetus, male fetus, female 2.3 yrs, female 2.4 yrs, female 24 yrs)
Abductor digiti minimi	1 (extra slips/tendons in female fetus L)/24 (normal in 10 male neonates, 6 neonates, male fetus, 5 female neonates, male 6 yrs)	0/26 (normal in 5 male neonates, 5 neonates, 2 male fetuses, 12 female neonates, 2 female fetuses)	0/7 (normal in female fetus, 2 male fetuses, female 2.3 yrs, female 2.4 yrs, female 24 yrs, female 29 yrs)
Flexor digitorum superficialis	6 (diminutive radial head and tendon to digit 5 in female neonate L, female neonate L, tendon 5 doubled in male neonate R, consisted of 4 separate fascicles in female neonate B, tendons 2 and 3 reversed in female neonate R, tendon 5 diminutive in male neonate B)/24 (normal in 8 male neonates, 6 neonates, male fetus, female neonate, male 6 yrs, female fetus)	11 (fused with fl. digitorum profundus to 5th digit in male neonate B, replaced by numerous muscles bellies in female neonate L, tendon 5 diminutive in female neonate L, female neonate B, female neonate B, male neonate B, male fetus B, male neonate L, female fetus B, tendon 5 doubled in male neonate R, extra fascicle supplying digiti quinti in female neonate B)/ 26 (normal in male neonate, 5 neonates, male fetus, 7 female neonates, female fetus)	2 (tendon 5 diminutive in male fetus B, double belly to fifth digit female 2.3 yrs R)/7 (normal in female fetus, male fetus, female 2.4 yrs, female 24 yrs, female 29 yrs)
Flexor digitorum profundus	5 (tendon/belly to digit 5 diminutive in male neonate R, female neonate L, tendon to digit 6 in male neonate L, consisted of 3 fascicles and the one closest to ulna supplied both 4th and 5th digits in female neonate L,	6 (fused with fl. digitorum superficialis to 5th digit in male neonate B, replaced by numerous muscles bellies in female neonate L, tendon/ belly to digit 5 diminutive in female neonate B, female neonate B, male fetus L, male	0/7 (normal in female fetus, 2 male fetuses, female 2.3 yrs, female 2.4 yrs, female 24 yrs, female 29 yrs)

Table 3 contd....

Table 3 contd.

Muscle anomalies	Trisomy 13 (presence versus absence, shown in bold)	Trisomy 18 (presence versus absence, shown in bold)	Trisomy 21 (presence versus absence, shown in bold)
Variant Muscles			
	does not pass through split in superficialis to 2nd digit in male fetus B)/**24** (normal in 8 male neonates, 6 neonates, 3 female neonates, male 6 yrs, female fetus)	neonate R)/**26** (normal in 3 male neonates, 5 neonates, male fetus, 9 female neonates, 2 female fetuses)	
Extensor digiti minimi	**2** (gives off extra tendons in female neonate L, split into 3 tendons distal to ext. retinaculum in male 6 yrs B)/**24** (normal in 10 male neonates, 6 neonates, male fetus, 4 female neonates, female fetus)	**12** (tendons displaced to radial or ulnar side in female neonate B, female neonate B, female neonate B, female neonate B, female neonate B, female neonate B, male neonate B, male neonate B, tendon doubled in female neonate L, male fetus B, female fetus B, doubled in female neonate R)/**26** (normal in 3 male neonates, 5 neonates, male fetus, 4 female neonates, female fetus)	**1** (tendon doubled in male fetus B (et quarti B))/**7** (normal in female fetus, male fetus, female 2.3 yrs, female 2.4 yrs, female 24 yrs, female 29 yrs)
Adductor pollicis	**2** (transverse head hypoplastic and arose from shaft of 4th metacarpal in male neonate B and male fetus B)/**24** (normal in 9 male neonates, 6 neonates, 5 female neonates, male 6 yrs, female fetus)	**0/26** (normal in 5 male neonates, 5 neonates, 2 male fetuses, 12 female neonates, 2 female fetuses)	**0/7** (normal in female fetus, 2 male fetuses, female 2.3 yrs, female 2.4 yrs, female 24 yrs, female 29 yrs)
Pronator teres	**4** (extra slip from fl. carpi radialis joined superficial head in male neonate R, doubled superficial head in male neonate R, arose partly from the fl. carpi radialis B and inserts onto radiocarpus on R in female neonate B, proximal head in female neonate B)/**24** (normal in 8 male neonates, 6 neonates, male fetus, 3 female neonates, male 6 yrs, female fetus)	**2** (missing ulnar head in female neonate L and female neonate R)/**26** (normal in 5 male neonates, 5 neonates, 2 male fetuses, 10 female neonates, 2 female fetuses)	**0/7** (normal in female fetus, 2 male fetuses, female 2.3 yrs, female 2.4 yrs, female 24 yrs, female 29 yrs)
Gluteus maximus	**1** (extra slips/tendons in male neonate R)/**20** (normal in 7 male neonates, 6 neonates, male fetus, 3 female neonates, male 6 yrs, female fetus)	**0/17** (normal in 2 male neonates, 5 neonates, male fetus, 8 female neonates, female fetus)	**0/5** (normal in female fetus, male fetus, female 2.3 yrs, female 2.4 yrs, female 24 yrs)

Table 3 contd....

Table 3 contd.

Muscle anomalies	Trisomy 13 (presence versus absence, shown in bold)	Trisomy 18 (presence versus absence, shown in bold)	Trisomy 21 (presence versus absence, shown in bold)
Variant Muscles			
Flexor digitorum brevis	**5** (lacked tendon to 5th toe in male neonate L, male neonate B, female fetus B, male fetus B, gave rise to only 2 tendons in male 6 yrs L)/**20** (normal in 6 male neonates, 6 neonates, 3 female neonates)	**0/17** (normal in 2 male neonates, 5 neonates, male fetus, 8 female neonates, female fetus)	**0/5** (normal in female fetus, male fetus, female 2.3 yrs, female 2.4 yrs, female 24 yrs)
Piriformis	**0/20** (normal in 8 male neonates, 6 neonates, male fetus, 3 female neonates, male 6 yrs, female fetus)	**0/17** (normal in 2 male neonates, 5 neonates, male fetus, 8 female neonates, female fetus)	**4** (split by peroneal nerve in female 2.3 yrs R, female fetus R, female 2.4 yrs B, Male fetus R)/**5** (normal in female 24 yrs)
Adductor longus	**1** (accessory head to head of tibia in female neonate B)/**20** (normal in 8 male neonates, 6 neonates, male fetus, 2 female neonates, male 6 yrs, female fetus)	**0/17** (normal in 2 male neonates, 5 neonates, male fetus, 8 female neonates, female fetus)	**0/5** (normal in female fetus, male fetus, female 2.3 yrs, female 2.4 yrs, female 24 yrs)
Semimembranosus	**2** (extra slips/tendons in male neonate R, male neonate B)/**20** (normal in 6 male neonates, 6 neonates, male fetus, 3 female neonates, male 6 yrs, female fetus)	**1** (extra slips between hamstrings in male fetus L)/**17** (normal in 2 male neonates, 5 neonates, 8 female neonates, female fetus)	**0/5** (normal in female fetus, male fetus, female 2.3 yrs, female 2.4 yrs, female 24 yrs)
Biceps femoris	**1** (extra slips/tendons in male neonate R)/**20** (normal in 7 male neonates, 6 neonates, male fetus, 3 female neonates, male 6 yrs, female fetus)	**1** (extra slips between hamstrings in male fetus L)/**17** (normal in 2 male neonates, 5 neonates, 8 female neonates, female fetus)	**0/5** (normal in female fetus, male fetus, female 2.3 yrs, female 2.4 yrs, female 24 yrs)
Gastrocnemius	**1** (extra slips/tendons in male neonate B)/**20** (normal in 7 male neonates, 6 neonates, male fetus, 3 female neonates, male 6 yrs, female fetus)	**0/17** (normal in 2 male neonates, 5 neonates, male fetus, 8 female neonates, female fetus)	**0/5** (normal in female fetus, male fetus, female 2.3 yrs, female 2.4 yrs, female 24 yrs)
Peroneus longus	**2** (extra slips/tendons in male neonate L, male neonate B)/**20** (normal in 6 male neonates, 6 neonates, male fetus, 3 female neonates, male 6 yrs, female fetus)	**0/17** (normal in 2 male neonates, 5 neonates, male fetus, 8 female neonates, female fetus)	**0/5** (normal in female fetus, male fetus, female 2.3 yrs, female 2.4 yrs, female 24 yrs)

Table 3 contd....

Table 3 contd.

Muscle anomalies	Trisomy 13 (presence versus absence, shown in bold)	Trisomy 18 (presence versus absence, shown in bold)	Trisomy 21 (presence versus absence, shown in bold)
Variant Muscles			
Plantaris	**2** (hypoplastic/lacks tendon in male neonate R, female fetus L)/**20** (normal in 7 male neonates, 6 neonates, male fetus, 3 female neonates, male 6 yrs)	**0/17** (normal in 2 male neonates, 5 neonates, male fetus, 8 female neonates, female fetus)	**0/5** (normal in female fetus, male fetus, female 2.3 yrs, female 2.4 yrs, female 24 yrs)
Quadratus plantae	**0/20** (normal in 8 male neonates, 6 neonates, male fetus, 3 female neonates, male 6 yrs, female fetus)	**2** (extra slip from tibia in male fetus R and female fetus B)/**17** (normal in 2 male neonates, 5 neonates, 8 female neonates)	**0/5** (normal in female fetus, male fetus, female 2.3 yrs, female 2.4 yrs, female 24 yrs)

CHAPTER 4

Cyclopia, Trisomic Anomalies, and Order *versus* Chaos in Development and Evolution

4.1 Introduction

The trisomy-associated anomalies reviewed here and described in the cyclopic Trisomy 18 fetus in Chapter 2 can be used to further our understanding of human development and evolution. In this and the following Chapters, we discuss the implications of our observations, comparisons, and literature review of muscle anomalies in Trisomies 18, 13, and 21 for developmental biology, evolutionary biology, and human medicine, with special attention to the theme of order versus randomness set out in Chapter 1.

4.2 Cyclopia and eye musculature

In the 28-week female fetus with Trisomy 18 and cyclopia we dissected, there was a single orbit with no metopic suture and no proboscis. The complete absence of a proboscis was peculiar, since most reported cases of cyclopia feature some structure above the eye resembling a nasal region. McGrath (1992) states "in cyclopia the face is as normal as the absence of median structures permits." The presence of a proboscis above the cyclopic eye is often assumed to be due to epigenetic factors, i.e., the nose cannot descend to its normal position because of the medial position of the eye. McGrath outlined the morphogenesis of the proboscis and how facial development is affected by the absence of median structures, but the complete absence of a proboscis in our case study may support a different model involving some other genetic or epigenetic perturbation, explaining why not all cyclopic cases result in the same morphology. This interpretation is consistent with Shapiro's theory of homeostasis, or random phenotypic effects (see Chapter 1 and below).

In the Trisomy 18 fetus we dissected, the "single" eye was revealed to have two fused irises. This fusion may have helped to produce the very wide orbit and **orbicularis oculi** with two semicircular portions. It has been theorized that the eye is an organizer of craniofacial development and may play a bigger role than previously thought in cyclopia (Kish et al., 2011). Early in normal development, the neural ectoderm induces initial frontal bone development (Schowing, 1968) in two distinct ossification centers. Could the single or partially fused eye, being an outpouching of the brain, contribute to abnormal signaling of frontal bone development? The missing metopic suture and fused frontal bone in this case study and various other reports (Mieden, 1982; Liu et al., 2011; Smith and Boulgakow, 1926) suggest that the eye or eyes might help to determine the morphology of the frontal bone, along with the brain. In support of this idea, the frontal bone's thickest regions, which may represent the ossification centers, were directly superior to the fused irises, paralleling their arrangement.

Previous reports on eye musculature in cyclopia have shown doubling of orbit muscles and bilateral absence of **rectus medialis** (e.g., Mall, 1917; Smith and Boulgakow, 1926). Our case is slightly different: the fused eyes lacked most medial muscles including rectus medialis and **obliquus superior**, but, peculiarly, there were two **obliquus inferior** muscles from which extra slips fused into new muscle bellies. Smith and Boulgakow (1926) reported a case of cyclopia, not related to trisomy, in which the obliquus inferior muscles crossed one another to insert on the midline of the fused eye but did not give off any extra muscle slips. The overall arrangement of head musculature found in the fetus we dissected could be due to the partially fused eyes, resulting in partial fusion of musculature close to the midline as well.

4.3 Development, trisomy, cyclopia, and muscles

Some variations in the cyclopic fetus we dissected could be attributed to developmental delay or early termination of muscle differentiation. The extra bellies connecting **flexor digitorum superficialis** and **flexor carpi radialis** to **flexor pollicis longus** might be lingering muscle slips resulting from failure of the flexor sheet to completely separate during development. Fusion of the head muscles **auricularis superior** and **auricularis anterior** might also be due to developmental delay. These muscles both arise from a common developmental anlage in the 2nd branchial arch, and delay very early in development could keep them together until later developmental stages (Gasser, 1967; Diogo and Abdala, 2010). Other cases of possible development delay will be discussed in Chapter 6, together with the role of evolutionary reversions and atavisms.

Other anomalies are harder to explain by developmental delay. The extra **biceps brachii** tendon inserting onto **pectoralis major** that we found bilaterally in the cyclopic Trisomy 18 fetus is hard to explain developmentally because these two muscles do not come from the same developmental anlage (Diogo and Abdala, 2010). This abnormality has been reported in some Trisomy 18 cases before: female neonate bilaterally (Diogo lab, 2013, unpublished data), female neonate bilaterally (Aziz, 1979), three neonate bilaterally (Aziz, 1981a,b), female neonate left (Barash et al., 1970), female neonate bilaterally, female neonate bilaterally, female neonate left, male neonate left (Ramirez-Castro and Bersu, 1978). Counting the bilateral finding in the Trisomy 18 cyclopic fetus we dissected, 11 of 28 (39%) Trisomy 18 individuals possessed this anomaly in at least one limb. Per limb, the ratio is 19 out of 56 (34%), or 11 of 28 (39%) left limbs and 7 of 28 (25%) right limbs (Barash et al., 1970; Colacino and Pettersen, 1978; Ramirez-Castro and Bersu, 1978; Aziz, 1979; 1980; 1981a,b; Pettersen, 1979; Pettersen et al., 1979; Bersu, 1980; Dunlap et al., 1986; Urban and Bersu 1987). Unilateral insertion of the biceps tendon onto pectoralis major on the right side has not been reported. The high (39%) incidence of this muscle defect in Trisomy 18 individuals studied so far combined with the 0% incidence in other trisomies for which there is some muscle data available (Trisomies 13 and 21; see Table 3) strongly contradicts

Shapiro's homeostasis theory of randomness and suggests that there is some "order" even in extreme human syndromes (Shapiro, 1983). However, these data do not necessarily support the broader implications and predictions of Alberch's "logic" theory because, according to this latter theory, one would expect to see the same birth defects in different syndromes (Alberch, 1989), as explained in Chapter 1 (see also below). To fully evaluate these competing ideas, one would need to dissect the musculature of more individuals with Trisomies 13 and 21 as well as with other syndromes and determine whether this specific muscle defect is also a frequent variation in karyotypically normal humans, as would be expected according to Alberch's "logic" theory.

Other anomalies were present both in the Trisomy 18 cyclopic fetus we dissected and in previous literature. Absence of **lumbricalis 1** was reported in five (18.5%) of 27 Trisomy 18 individuals (including ours), 0 (0%) of seven Trisomy 21 individuals, and three (13%) of 24 Trisomy 13 individuals (Barash et al., 1970; Colacino and Pettersen, 1978; Ramirez-Castro and Bersu, 1978; Aziz, 1979; 1980; 1981a,b; Pettersen, 1979; Pettersen et al., 1979; Bersu, 1980; Dunlap et al., 1986; Urban and Bersu, 1987). Doubling of **extensor indicis** (two tendons) was reported in 12 (44.4%) of 27 Trisomy 18 individuals (including ours), two (28.6%) of seven Trisomy 21 individuals, and six (25%) of 24 Trisomy 13 individuals (Barash et al., 1970; Aziz, 1979; 1980; 1981a,b; Bersu, 1980; Colacino and Pettersen, 1978; Dunlap et al., 1986; Pettersen, 1979; Pettersen et al., 1979; Ramirez-Castro and Bersu, 1978; Urban and Bersu, 1987) (Table 3). These anomalies provide some evidence for the presence of the same defect in more than one trisomy syndrome, contradicting Shapiro's "randomness" theory and supporting Alberch's "logic" theory. However, further study should still be undertaken since these percentages refer to a still relative small sample.

Multiple anomalies previously reported in Trisomy 18 did not appear in the Trisomy 18 cyclopic fetus we dissected. We were unable to determine if **platysma cervicale** (a distinct bundle of the platysma complex running from the region of the mouth to the nuchal region that is often absent in adult humans but often present in adult non-human mammals) was present due to removal of some parts of the head musculature prior to our dissections; this muscle was reported in 13 (76.5%) of 17 Trisomy 18 individuals, five (100%) of five Trisomy 21 individuals and five (25%) of 20 Trisomy 13 individuals (Barash et al., 1970; Bersu and Ramirez-Castro, 1977; Bersu, 1980; Colacino and Pettersen, 1978; Aziz, 1979; 1980; 1981a,b; Pettersen, 1979; Pettersen et al., 1979; Dunlap et al., 1986; Urban and Bersu, 1987) (Table 3). **Rhomboideus occipitalis** (a muscle that runs from the superior angle of the scapula to the occipital bone of the skull, usually not present in karyotypically normal humans) was reported in eight (29.6%) of 27 Trisomy 18 individuals, 0 (0%) of seven Trisomy 21 individuals and one (4%) of 24 Trisomy 13 individuals (Barash et al., 1970; Bersu and Ramirez-Castro, 1977; Aziz, 1979; 1981a,b; Dunlap et al., 1986; Urban and Bersu, 1987), but it was not present in the fetus we dissected. While the presence of platysma cervicale can be attributed to a delay in development (Gasser, 1967; see above), rhomboideus occipitalis cannot because there is no evidence that this muscle is usually present during early developmental stages of karyotypically normal humans.

A common theory concerning muscular abnormalities holds that more variant structures in the karyotypically normal human population are the first structures to be lost in disorders such as trisomies: "the characteristics most frequently abnormal [in Down syndrome] were those that had been considered developmentally less stable in the general population" (Shapiro, 1983). This aspect is one of the few shared by the otherwise opposing theories of Alberch and Shapiro: both predict that many birth defects, such as the absence of a certain muscle, also appear as frequent variations in the general population. For example, **palmaris longus** is missing on at least one side of the body both in a great majority of individuals with trisomies (Trisomy 13 (83%), 18 (88.4%), and 21 (85.7%)) and in a large minority of the karyotypically normal human population (10 to 20%; Diogo and Wood, 2012a) (Table 3). In the fetus we dissected, palmaris longus was completely absent on the left but intact on the right. This asymmetry lends further support to the theory of homeostasis and more random phenotypic effects. **Palmaris brevis** is often absent in Trisomy 13 (83% individuals), 18 (58% individuals), and 21 (14.2% individuals) (Table 3). This muscle was also missing bilaterally in the fetus we dissected, lending support to both the "logic" and "homeostasis" theories, because the palmaris brevis is often variable in the general human population and has multiple possible attachments (Diogo and Wood, 2012a).

4.4 Integration and limb serial homology (Fig. 4.1)

In the Trisomy 18 cyclopic fetus we dissected, we found evidence of a common pattern in the structures of the short flexors and extensors of the hands and feet. In both hands, **extensor digitorum** and **flexor digitorum superficialis** were missing their 4th tendons to the digit 5. In both feet, **flexor digitorum brevis** was missing its 4th tendon to the digit 5, and its 3rd tendon to digit 4 was very diminutive. In the left foot, the arrangement of the extensors was normal, while in the right foot **extensor digitorum longus** was missing its 4th tendon to digit 5, and **fibularis tertius** had a separate muscle belly. However, fibularis tertius and the 3rd tendon to digit 4 of extensor digitorum longus sent tendinous branches to one another, resulting in a small branch of tendon extending along fibularis tertius and inserting onto the distal phalanx of digit 5. In summary, both the flexors and extensors in the hands and feet were missing a tendon to digit 5, except for extensor digitorum longus in the left foot (Fig. 4.1).

Of the reports of missing flexor or extensor tendons in Trisomies 13 and 18, a missing flexor digitorum superficialis tendon to digit 5 was reported most frequently, in nine (53%) of 17 Trisomy 13 and 18 cases and 0 (0%) of seven Trisomy 21 cases (Ramirez-Castro and Bersu, 1978; Pettersen, 1979; Dunlap et al., 1986; Urban and Bersu, 1987). However, only Trisomy 18 individuals [five (56%) of these nine] were missing both the tendons to digit 5 of the flexor digitorum longus and brevis of the foot (Ramirez-Castro and Bersu, 1978). In two of seven Trisomy 13 cases, the same hand and foot integration was also apparent (Pettersen, 1979; Pettersen et al., 1979). Many cases investigated for limb variations in Trisomies 13, 18, and 21 do not include the lower limbs, so these percentages could vary greatly (Dunlap et al., 1986). These numbers do not reflect the limb musculature in cases of human cyclopia because there has been no previous detailed report on the limb muscles in cyclopic individuals. The missing tendons of the extensors and flexors in the fetus we dissected and other reports of trisomies without cyclopia suggest developmental integration between digital flexors and some extensors in upper and lower limbs, lending support to the theory that hands and feet share similar developmental mechanisms. Our study of humans cannot answer the question of whether these mechanisms are shared due to an original duplication and thus similarity of the pectoral and pelvic appendages in early gnathostomes or instead to derived co-option of similar genes in the transitions from fins and limbs, as recently argued by Diogo et al. (2013a) (see Chapter 1). However, the similarities between these leg/flexor muscles in Trisomies 13 and 18 do strongly support Alberch's "logic" theory, as will be discussed in Section 4.6.

4.5 Facial muscles and topological position *versus* developmental anlage in the cyclopic head

In Chapter 1, we hypothesized that, in the Trisomy 18 cyclopic fetus as well as other trisomic individuals, we would see more defects associated with a change in the bony attachments of muscles in the limbs than in the head. This hypothesis was based on Köntges and Lumsden's (1996) model of the involvement of neural crest cells in the attachments of head muscles onto the branchial arches from which head muscles derive. The normal phenotype shows a few exceptions to Köntges and Lumsden's (1996) model, including the mylohyoideus, a muscle of the 1st arch that inserts onto the mandible but also often has a small insertion onto a derivative of the second (hyoid) arch, the hyoid bone. However, as noted by Köntges and Lumsden's (1996) most head muscles insert onto skeletal structures derived from their own arch. In the Trisomy 18 cyclopic fetus, the observable branchiomeric muscles (i.e., muscles of the branchial arches) other than the muscles of facial expression usually attached to the bones/structures to which they normally attach, despite abnormal skeletal structure. An additional mandibular (i.e., innervated by cranial nerve V) muscle was present, the **intermandibularis anterior** (Fig. 2.8), but the attachments of this muscle were similar to those usually seen in the non-human taxa in which the muscle normally occurs (between the two mandibles). However, the muscles of facial expression (derived from the 2nd branchial arch and innervated by the facial nerve, as are the other hyoid muscles: stylohyoideus, digastricus posterior, and stapedius) showed more dramatic changes in their attachments. For instance, we predicted that the nasal musculature would be missing or displaced due to the absent/not ventrally migrated nasal bones and cartilage, but this was not the case. A possible **nasalis** muscle was identified spanning the superior portion of the maxilla just below the orbit where the nose normally lies; its origins were on the lateral maxilla, and the fibers converged to insert along the midline prominence just below the orbit (Fig. 2.11).

Unlike other branchiomeric muscles, the muscles of facial expression share with the limb and hypobranchial (i.e., infrahyoid and tongue) muscles expression of a gene (*Met*) that is involved in the migration of the muscles to their final positions (Prunotto et al., 2004). Accordingly, the muscles of facial expression do not normally attach onto skeletal structures of the 2nd branchial (hyoid) arch, but instead onto other skeletal structures and/ or directly onto skin. The observations, comparisons and literature review provided in this book suggest that the facial muscles share even more similar developmental mechanisms with the limb muscles than previously assumed, because their attachments seem to be mainly directed by the topological positions of skeletal structures. The insertion of the presumptive nasalis in the region where the nose would normally be is a powerful illustration of this idea (Fig. 2.11). In contrast, muscles of the head other than those of facial expression, such as the masseter and temporalis (which derive from the 1st, or mandibular, branchial arch) attached normally to structures derived from the same arch.

We can therefore speculate that topological position directed all muscle attachments in early chordate evolution, because non-chordate animals have no neural crest cells (Gilbert, 2000; N.B., there is still controversy about whether non-vertebrate chordates such as urochordates have neural crest-like cells: see, e.g., Hall and Gillis, 2013). This basal "muscle attachment-topology link" seems to persist in the appendicular muscles of fish and tetrapods, as will be explained in Chapter 5. However, during the transitions that led to vertebrates, development of the head may have caused muscles in this area to become more constrained, as control of head muscle attachments by neural crest cells increased anatomical consistency within the very complex and roughly spherical structure of the head. Then, during the transitions to mammals, muscles of facial expression evolved by modifications/expansions of existing muscles of the second (hyoid) branchial arch (Diogo and Wood, 2012a), and they seem to have reverted to the more basal topological position-type of attachments seen in the limbs muscles, thus explaining the developmental similarities shared by facial and limb muscles. This scenario would help to explain the great diversification of facial muscles and their attachments, and therefore of facial expressions, seen in mammals. It would also help explain why there is so much variation in facial musculature configuration and attachments in primates and in the karyotypically normal human population (Diogo and Wood, 2012a).

The possible nasalis muscle in the cyclopic fetus we dissected has important implications for our understanding of cyclopia. The record of facial muscles in previous literature on cyclopia is minimal, and those who did examine the face stated that the musculature was too fused to permit recognition of specific muscles (Mieden, 1982). Still, definitive absence of nasal muscles has not been reported either, so the nasal muscles may have been included within the fused facial muscle masses. Cyclopic human individuals usually have a proboscis above the eye, but the one we examined did not, making comparisons with the previous literature even more difficult. In literature on the anatomy of the proboscis in human cyclopia, structures representing nasal cartilages, ethmoidal nervous structures and blood vessels, the conchae, olfactory mucosa, respiratory mucosa, and nasolacrimal duct have all been reported within the proboscis (McGrath, 1992; Hill, 1920). However, these reviews did not mention nasal musculature within the proboscis, reinforcing the idea that the facial muscles that normally attach to the nose in the normal population do not insert in those skeletal structures in individuals with cyclopia. Our theory of a "nearest neighbor" topology-mode of attachment for the facial muscles predicts that, when the proboscis is present in cyclopia, the true nasal muscles are not attached to it, but instead are attached to the normal region of the nose, as was the case in the cyclopic fetus we dissected.

4.6 "Logic of monsters", homeostasis, and order *versus* chaos in development and evolution

None of the previous publications on muscle defects in Trisomies 13, 18, and/or 21 has compiled information about these defects in a systematic way. We do so in this book to make it easier to understand the general frequency and patterns of muscle defects in these trisomies and their medical and theoretical implications. Mainly based on their own dissections, observations, and comparisons, some authors have proposed hypotheses to explain the frequent occurrence of muscle defects, and additional muscles in particular, in individuals with these trisomies. For instance, Opitz and colleagues (e.g., Barash et al., 1970; Opitz, 1985; Opitz and Gilbert-Barness, 1990) proposed a "development field concept" hypothesis and defended the idea that the occurrence of muscle defects in Trisomies 13, 18, and 21 reflects not only incomplete but also abnormal differentiation of muscles.

A broader theory was proposed by Shapiro (1983) regarding Down syndrome and the effects of trisomy in general (see also Chapter 1). According to this theory, trisomies are generally associated with decreased developmental and physiological homeostasis, resulting in greater displacement from 'normal' developmental pathways and physiological reactions. The presence of a whole extra functioning chromosome or large chromosome segment causes generalized disruption of evolved genetic balance. Because of the obligatory integration of the entire genotype, this disruption affects not only products of the trisomic chromosome but other chromosomes as well. The disruption of evolved balance results in decreased developmental and physiological buffering against genetic and environmental forces, leading to decreased homeostasis. Developmental pathways and physiological processes vary in their stability depending on the degree of canalization or precision of homeostatic controls affecting them. The more unstable pathways and processes will be most often and most seriously affected in the trisomic phenotype. Specifically, this theory predicts that generalized diminished buffering capacity results in increased variance for metric traits, amplified instability of developmental pathways, reduced precision of physiologic homeostatic controls, and increased susceptibility to genetic and environmental trauma.

A different, even broader theory was proposed by Alberch (1989) concerning not only trisomies but in fact birth defects and variations in general. Based on his own studies of various tetrapod taxa and an extensive review of the literature, this theory, unfortunately designated the 'logic of monsters', states that ontogenetic constraints are so strong that a) even conditions caused by abnormalities in different genes or chromosomes ultimately lead to similar musculoskeletal defects, and b) the anatomical (including muscle) defects more commonly found in human syndromes (such as those associated with Trisomies 13, 18, and/or 21) are phenotypically similar to muscle anomalies/variants found in karyotypically normal humans.

The data compiled in the present book provide some evidence supporting of both the "logic"/order and the "homeostasis"/random theories of birth defects (see Chapters 2 and 3), but there is more support for the former. While individuals with different trisomies shared many similar anomalies, some were particular to Trisomies 13, 18, and/or 21. For example, the supernumerary muscle **platysma cervicale** is common across all three trisomies, but the presence of a **rhomboideus occipitalis** and the fused deltopectoral complex are most common in only one or two trisomies (see Table 3). Further study of these anomalies and their relationship to the genotypes of trisomy syndromes would help to explain why some defects are shared across trisomies and others are not.

Another example that seems to provide support for the "logic" theory is the configuration of the long extensors and flexors of the hands and feet of the cyclopic fetus we dissected (Fig. 4.1). As explained above, in both hands, the **extensor digitorum** as well as the **flexor digitorum superficialis** were all missing their 4th tendon to the digit 5. In both feet, the **flexor digitorum brevis** was missing its 4th tendon to the digit 5, and its 3rd tendon to digit 4 was very diminutive. In the left foot, the arrangement of the extensors was normal, while in the right foot the **extensor digitorum longus** was missing its 4th tendon to digit 5 and the **fibularis tertius** was a separate muscle. However, the fibularis tertius and the 3rd tendon to digit 4 of the extensor digitorum longus sent tendinous branches into one another, resulting in a small branch of tendon extending along the fibularis tertius and inserting onto the distal phalanx of digit 5. At first glance, the configuration of the muscles in different limbs and parts of the limbs (the flexor versus extensor musculature) seems chaotic. However, there is a strikingly similar and clearly non-random overall pattern: except for the extensor digitorum longus of the left foot, flexors and extensors of both hands and feet are missing a tendon to digit 5 (Fig. 4.1).

Some observations support both the "logic" and the "homeostasis" hypotheses, e.g., both predict that trisomic individuals will have a high frequency of abnormalities in structures that are considered to be developmentally less stable, and thus more variable, in the general population, such as the **palmaris longus** (see Chapter 3). Lastly, some evidence supports the "homeostasis" theory but not the "logic" theory, including asymmetrical birth defects in Trisomies 18, 13, and 21 (see Chapters 2 and 3). More detailed studies will be required make a compelling argument in support of one or the other hypothesis, thus answering a crucial evolutionary and developmental question.

CHAPTER 5

Digits and Muscles: Topology-Directed Muscle Attachment

5.1 Introduction

Evolution of limbs in tetrapods enabled the spectacular transition from water to land habitats (Romer, 1933; Bowler, 1996; 2007; Wagner and Chiu, 2001; Fabrezi et al., 2007; Wagner and Larsson, 2007; Weatherbee and Niswander, 2007; Laurin, 2011). A major goal of tetrapod research is to explain the patterns, causes, and consequences of the considerable morphological variation in the forelimbs and hindlimbs of tetrapods, both within and diverging from the pentadactyl bauplan of extant tetrapods (N.B., ancestral tetrapods such as *Ichthyostega* and *Acanthostega* had up to eight digits: Coates and Clack, 1990). Developmental studies can provide some insight into these patterns by explaining the mechanisms upon which evolution can act. While somatic limb muscle progenitor cells apparently do not carry intrinsic positional information (Duprez, 2002), condensations that will give rise to bones may provide the positional signaling for the subsequent development of muscles (Manzano et al., 2013). Muscles can also play an important role in at least some aspects of skeletal morphogenesis. For example, muscle contraction may help to regulate chondrocyte intercalation and skeletal elongation, thus facilitating coordination between muscle and skeletal development (Shwartz et al., 2012).

Information obtained from non-pentadactyl limbs is crucial to clarify how the functional and spatial associations between bones and muscles have changed during the evolution of morphological variation in limbs, as well as the development of the common limb birth defects found in humans and other species (Dunlap, 1967; Muntz, 1975; Shubin and Alberch, 1986; Kardon, 1998; Wagner and Chiu, 2001; Coates et al., 2002; Duprez, 2002; Wagner and Larsson, 2007; Weatherbee and Niswander, 2007; Ponssa et al., 2010; Laurin, 2011; Shwartz et al., 2012; Manzano et al., 2013). However, myological data on non-pentadactyl limbs are relatively rare in the literature. The scarcity of such information is surprising because limb reduction has long attracted researchers' attention (e.g., Owen, 1849; Presch, 1975; Caputo et al., 1995; Diogo et al., 2013b) for its potential to elucidate broader evolutionary themes, such as evolutionary trends, anatomical convergence, and evolutionary reversals that violate "Dollo's law" (Diogo and Abdala, 2010; Diogo and

Wood, 2012a; Diogo and Tanaka, 2014). Inferences of evolutionary patterns in limbs and the study of human limb birth defects are usually based on either external morphology or osteology, frequently ignoring muscular anatomy. Therefore, because anatomical and developmental studies usually focus on pentadactyl autopodia (hands/feet) and those dealing with deviations from the norm focus on reduction or gain of cartilages and/or bones, the muscular anatomy of non-pentadactyl limbs remains unexplored.

The scarcity of data on the relationship between hard and soft tissues in non-pentadactyl limbs affects not only the knowledge of broad evolutionary subjects but also human evolution and human medicine (Diogo and Tanaka, 2014). Indeed, changes in the number of digits are the most common anomalies of human limbs at birth (e.g., the presence of an extra toe and/or finger has a 0.2% incidence, i.e., 1 in 500 births), but information about the soft tissue changes that occur in these anomalies is extremely scarce (Castilla et al., 1996). Non-pentadactyly, especially when it concerns the complete duplication of, or failure to form, a digit, profoundly affects the functioning of the limbs, especially the hands, which are used for complex and fine tasks in daily life (Waters and Bae, 2012). Surgery is therefore often recommended and performed within the first years of life to improve the biomechanical function of the limb and to provide the most natural look, feel, and function of the corrected limb for the infant or child (Watt and Chung, 2009). Surgical options depend on the specific type of defect, and the level of duplication/reduction of both the hard and soft tissues guides the operative treatment of poly- and oligodactyly (Tonkin and Bulstrode, 2007). Therefore, such surgeries are often very complex and, due to the scarcity of muscle studies, the specific muscle configuration and attachments found in such limbs is poorly known and thus difficult to predict, particularly in the less studied types of defects (Waters and Bae, 2012).

Many limb anomalies are attributable to homeotic transformation (replacement of a normal body part by one that normally forms in another region of the body)—one of the most important current topics in evolutionary developmental biology. For instance, in preaxial polydactyly, one of the most common congenital anomalies of the human hand, the duplication of the thumb gives the two most radial digits the homeotic identity of digit 1 (Castilla et al., 1996). Homeotic transformations have also played an important role in the evolution of normal phenotypes. For example, it is now commonly accepted that the digits of the adult bird wing derive from the second, third and fourth developmental anlagen (embryonic condensations), but that homeotically and morphologically these digits correspond to digits 1, 2 and 3 of other tetrapods. A similar homeotic transformation seems to have occurred in the hand of the three-toed Italian skink *Chalcides chalcides* (Young et al., 2009).

In order to study the spatial associations between limb bones and muscles, we have recently completed a long term, multidisciplinary project (Diogo and Abdala, 2010; Diogo and Wood, 2012a; Diogo and Tanaka, 2014; Diogo and Molnar, 2014) describing and comparing the appendicular muscles of all major groups of tetrapods in order to reconstruct their evolution. The project included dissections and developmental and regenerative studies. *The data reported in this book thus allow us to combine the results of that project with our new investigations of how human birth defects involving the formation of non-pentadactyl limbs influence muscle attachments.* By doing so, we were able to test whether the anatomical patterns seen in birth defects follow those seen in wild-type non-human tetrapods with non-pentadactyl limbs, and therefore whether the study of birth defects could reciprocally illuminate evolutionary patterns. By providing a broad discussion of regenerative, developmental, comparative, and teratological works, this project paves the way for future developmental experimental and mechanistic studies with crucial implications for evolutionary and developmental biology and human medicine. The information provided below is based on a paper published with our colleagues John Hutchinson and Virginia Abdala and one of our former undergraduate students, Sean Walsh, and we thank them for their collaboration, wise insights, and friendship (Diogo et al., in press).

5.2 Tetrapod limbs, digits, muscles, and homeotic transformations

Each digit has a specific topological position, anlage number, and homeotic identity. Topological position refers to the adult relationship with other structures and to adult spatial data, not necessarily the position of the developmental anlagen. For instance, the topological position of the adult avian digit that derives from the second developmental anlage is digit 1, because this is the most radial (medial/preaxial) digit in the adult. In this case, the topological position (digit 1) and homeotic identity (digit 1) are the same and are different from the developmental anlage from which the digit develops (the second anlage) (Young et al., 2009).

The combination of analyses of wild-type non-human tetrapods and human birth defects provides a unique perspective on the long-standing evolutionary question raised by authors such as Owen (1849): do tetrapods share a general, predictable spatial correlation between limb bones and muscles? We found a surprisingly consistent pattern, both in the non-pentadactyl limbs of wild-type taxa such as frogs (Diogo and Ziermann, 2014), salamanders and chickens, and in the humans with birth defects: the identity and attachments of the forelimb and hindlimb muscles are mainly related to the physical (topological) position, and not the number of the anlage or even the homeotic identity, of the digits to which the muscles are attached. For instance, the Tables included in one of our recent papers (Diogo and Molnar, 2014) provide numerous examples in which loss of digits in wild-type tetrapods is accompanied by change in the insertions of muscles to the adjacent remaining digit. As an example, the hand digits of urodeles (salamanders) such as axolotls derive from the anlagen, and have a homeotic identity, of digits 1–4 (Young et al., 2009), but digit 4 is the attachment point of the **abductor digiti minimi** muscle that usually goes to digit 5 in pentadactyl tetrapods.

As predicted by our hypothesis, the **abductor pollicis brevis** in birds, where present, attaches to the most radial digit (derived from the 2nd anlage), whereas in pentadactyl taxa this muscle always inserts onto digit 1, which derives from the first, and not the second, anlage (Diogo and Molnar, 2014). Another example concerns the most ulnar (lateral/axial) digit of the hand in the above-mentioned axolotls, topologically similar to digit 5 in pentadactyl tetrapods, develops from the anlage of digit 4 and has a homeotic identity of digit 4. This case supports our hypothesis because muscles that normally go to digit 5 in pentadactyl tetrapods, such as the abductor digiti minimi, insert on the most ulnar digit despite its developmental origin and homeotic identity as digit 4. Similarly, the hand digits of frogs are derived from the anlagen, and have a homeotic identity, of digits 2–5 (Young et al., 2009), but the **abductor pollicis longus** in frogs develops ontogenetically in connection with digit 2 exactly as it develops in connection with digit 1 in pentadactyl tetrapods—i.e., lying radial and somewhat deep to the extensor digitorum and running distoradially to attach onto metacarpal II/digit 2 (Diogo and Molnar, 2014). The detailed study of non-pentadactyl tetrapod feet also consistently supports our hypothesis; for example, in taxa such as crocodylians and birds that have only digits 1, 2, 3 and 4 in the foot, digit 4 is the insertion site for the abductor digiti minimi, which in pentadactyl feet always goes to digit 5 (Diogo and Molnar, 2014).

These observations also are consistent with evolutionary and biomechanical theory, because the most pre-/post-axially extreme digits (i.e., digits 1 and 5) are often specialized anatomically for increased mobility and/or are moved by individual muscle, such as the abductor pollicis longus and the abductor digiti minimi. Biomechanical function explains why loss of, for example, digit 1 in the forelimb of birds is accompanied by a homeotic transformation in which the most radial digit of the wing (derived from the anlage of digit 2) recovers the identity of digit 1. Even in cases of digit reduction where there is no homeotic transformation, developmental mechanisms (configuration/identity of muscles related to position, and not identity, of digits) ensure that the extremities retain the particular muscles related to the specialized functions of digits 1 and/or 5.

5.3 Birth defects, limb muscles, non-pentadactyly, and implications for human medicine

To ascertain whether patterns in the non-pentadactyl autopodia of wild-type tetrapod taxa are also consistently found in cases of birth defects, we dissected humans with non-pentadactyl hands, some of which are described in this book. Because it is the first detailed study of both the soft and hard tissues of non-pentadactyl human autopodia from individuals with different genetic backgrounds, this project allowed us to establish correlations between the observed patterns and birth defects with specific genetic conditions. The predictions from our topology-based attachment hypothesis were consistently borne out by our dissections, even in the most extreme cases, such the Trisomy 18 human newborn (Figs. 1.4, 1.3) with four digits on one of the hands (thumb missing) and six digits on the other hand (two thumbs present). In the hand with four digits, all muscles that normally insert onto the thumb were present but attached instead onto digit 2 (Fig. 1.4), while in the hand with six digits the muscles that normally attach onto the radial (e.g., **adductor pollicis brevis**) and ulnar (e.g., **adductor pollicis**) sides of the thumb attached respectively onto the radial and ulnar sides of the more radial and more ulnar thumbs (Fig. 1.3). Our determination of the identity and homology of the muscles in both normal and abnormal phenotypes was based not only on topology, which could lead to circular reasoning, but on all available data (e.g., innervation, orientation of the fibers, origin, insertion, divisions, topological relationship to other muscles and bones). For example, the muscle identified as adductor pollicis in Fig. 1.3 is innervated by the deep ulnar nerve, has disto-radially directed fibers, originates from the contrahens fascia, and is divided into oblique and transverse heads, all of which characterize the adductor pollicis of the normal human phenotype. The only difference was that, in the hand shown in Fig. 1.3, adductor pollicis inserts on digit 2 and not the thumb.

Other cases like the hand with six digits, which generally support our hypothesis, have been described in the scarce literature on the associations between soft and hard tissues in non-pentadactyl human limbs (e.g., Light, 1992). In one case of a hand with six digits, designated preaxial polydactyly, the duplication of the thumb (giving the two most radial digits the homeotic identity of digit 1) was not accompanied by duplication of the muscles that normally go to the thumb. Instead, the muscles abductor pollicis brevis and **flexor pollicis brevis**, which are the most radial thumb muscles in the normal phenotype, attached to the more radial of the two thumbs, and the adductor pollicis, which is normally the most ulnar thumb muscle, attached to the more ulnar of the two thumbs. In other words, the muscles were not simply duplicated, as were the thumb bones, but instead attached to one or the other thumb according to the adult topological position of each of the duplicated digits.

Our studies of human birth defects also support Alberch's (1989) theory that developmental or morphological constraints are so important that a predictable, logical pattern exists even in cases of extreme anatomical defects which often mirrors the patterns seen in the normal phenotype of other taxa. However, there are a few exceptions to this rule. For instance, Heiss (1957) described a peculiar case of a human subject with two pentadactyl hands and no thumbs in which, contrary to the cases cited above, all normal thumb muscles were missing and there were no major topological changes to other muscles. This configuration seems to characterize the rare human disorder "tri-phalangeal thumb", a malformation of digit I involving perfect homeotic transformation of the thumb into an index finger in which the muscles normally associated with the thumb are absent (e.g., abductor/opponens/adductor pollicis) (Heiss, 1957). If future dissections of humans with "tri-phalangeal thumbs" confirm these reports, this syndrome might constitute an exception to our hypothesis in which genetic and/or epigenetic factors cause the identity and attachments of the muscles to follow to the homeotic identity of the digits to which they attach rather than their topological position. This exception would imply hitherto-unrecognized developmental plasticity in the patterning of hard and soft tissues and thus open up new lines of inquiry in the field of developmental biology. However, we did not find any similar exceptions in any of the non-pentadactyl limbs (both hands and feet) of the wild-type tetrapods studied (Diogo and Molnar, 2014; see above), suggesting that such plasticity would more likely be the result of extreme defects than true developmental plasticity seen during normal ontogeny in tetrapods.

In summary, this new comparative synthesis addresses important questions in developmental and evolutionary biology and human medicine and paves the way to future mechanistic developmental studies of the ontogeny and specific genetic and/or epigenetic causes of certain developmental and evolutionary patterns. For instance, a recent study showed that the cells that give rise to the limb bone eminences to which the muscles attach are descendants of a unique set of *Scx*- and *Sox9*-positive progenitors, and that these bony eminences emerge only after the primary cartilage rudiments have formed (Blitz et al., 2013). The cells that give rise to these eminences are not descendants of chondrocytes, and the formation of bony eminences is therefore external to and independent of the formation of the developing bone and are added to it in a modular fashion (Blitz et al., 2013). This developmental modularity might explain some of the patterns we have observed in non-pentadactyl limbs, such as muscles that normally attach to digits 1 and/or 5 in pentadactyl taxa attaching to digits 2 and/or 4 instead. This hypothesis could be tested experimentally by blocking the expression of *Scx* and *Sox9* (e.g., Blitz et al., 2013) in developing non-pentadactyl limbs to see whether this intervention interferes with the modular pattern of change in muscle attachments. Such experimental and mechanistic studies have the potential to clarify the developmental processes that produce the anatomical patterns described here and specifically test the hypothesis that muscle identity/attachment is mainly related to digit topology. Detailed knowledge of these patterns and processes, and their possible exceptions, would also be of great value to physicians and surgeons because they help to explain the configuration of soft tissues in the non-pentadactyl limbs so commonly found in human birth defects. Because the pattern we describe (Figs. 1.3, 1.4) is not always found in humans with non-pentadactyl limbs (see, e.g., reference to Heiss, 1957; above), further studies on humans with birth defects are needed to identify specific cases/syndromes/defects that are exceptions. These data could improve surgical outcomes and reduce the number of surgeries needed, thus benefitting patients, clinicians, surgeons, and the medical and scientific community in general.

CHAPTER **6**

Evolutionary Mechanisms and Mouse Models for Down Syndrome

6.1 Introduction

Our observations and comparisons provide a fresh perspective on one of the most controversial subjects in developmental and evolutionary biology: the notion of *atavism* and the related concepts of evolutionary reversions and Dollo's law. Dollo's Law states that once a complex structure is lost it is unlikely to be reacquired (e.g., Gould, 1977; 2002). Atavism is the evolutionary reversion of a trait to its ancestral state, which may violate Dollo's law if it involves regaining a complex structure. In a recent paper, Wiens (2011; see, e.g., his Table 2) listed several examples of violations of Dollo's Law, including the loss of mandibular teeth in the ancestor of modern frogs > 230 million years (MY) ago and their reappearance in the frog genus *Gastrotheca* during the last 5 to 17 MY, the reappearance of a larval stage in plethodontid salamanders and in hemiphractid frogs, the reappearance of digits in several clades of lizards (amphisbaeninan lizards of the genus *Bipes*, gymnophthalmid lizards of the genera *Tretioscincus* and *Bachia*, and scincid lizards of the genus *Scelotes*), the reappearance of eggshell in the boid snake *Eryx*, and the reappearance of shell coiling in snails. However, none of the examples cited in Wiens's survey involves muscles; like most studies on the evolution of tetrapods and other vertebrates, it was based on hard tissue data.

To address the paucity of data on soft tissues such as muscle, we recently reported the results of a long-term study of the comparative anatomy, homologies, and evolution of the head, neck, pectoral and forelimb muscles of all major groups of vertebrates based on dissection of hundreds of specimens and a review of the literature (Diogo et al., 2008a; 2009a,b; 2010; Diogo and Abdala, 2010; Diogo and Wood, 2011; 2012a,b). Diogo and Wood (2011) combined data from their dissections with carefully validated information from the literature to undertake the first comprehensive parsimony and Bayesian cladistic analyses of the order Primates based on myological data for each of the major primate higher taxa and for a range of outgroups (tree-shrews, dermopterans, and rodents). The most parsimonious tree obtained from the cladistic analysis of 166 characters taken from the head, neck, pectoral and upper limb musculature, shown in Fig. 6.1, is fully congruent with Arnold et al.'s (2010) evolutionary molecular tree of Primates and similar to the primate molecular trees obtained by Fabre et al. (2009) and Perelman et al. (2011), with the exception that the two latter studies did not recover the Cebidae as a monophyletic group. The results of Diogo and Wood (2011), as well the few other cladistic analyses based on soft tissues published to date, reveal that soft tissues can be particularly useful for inferring phylogenetic relationships, including those among fossil taxa such as dinosaurs. When homoplasy of muscles and hard tissues are directly compared, the former tend to be less homoplastic than the latter (for a recent review, see Diogo and Abdala, 2010). In addition, the inclusion of soft tissue-based information in phylogenetic investigations allows researchers to address evolutionary questions that are not tractable using other types of evidence, including the evolution, functional morphology, homoplasy (including evolutionary reversions), and neotenic features of our own clade and closest living relatives (Diogo and Wood, 2012a,b). In this Chapter, we combine data from our myology-based cladistic analyses of evolutionary reversions that do or do not violate Dollo's law within the primate clade with our observations and literature review on muscle birth defects. We explore the implications of these comparative and phylogenetic studies for the understanding of the evolution, ontogeny, and variability of primates and modern humans, particularly the importance of reversions in primate and human evolutionary history and in birth defects.

6.2 Evolutionary reversions, Dollo's law, and human evolution

The results of Diogo and Wood (2012a,b) concerning evolutionary reversions in primates and their closest living relatives are summarized in the tree in Fig. 6.1. These results suggest that evolutionary reversions played a substantial role in primate and human evolution, at least with respect to the musculature of the head and neck, pectoral region, and upper limb. One in seven of the 220 evolutionary transitions that are unambiguously optimized in this tree is a reversion to a plesiomorphic state (N = 28). Of the 28 reversions shown in this tree, six occurred at nodes that led to the origin of modern humans (Fig. 6.1). One occurred at the node leading to the Hominoidea (q': reversion of "**Temporalis** has a pars suprazygomatica"), one at the node leading to the Homininae (t': reversion of "Latissimus dorsi and teres major are fused"), and four occurred within the sub-tribe Hominina (y', z', α' and β': reversions of "Anterior portion of **sternothyroideus** extends anteriorly to the posterior portion of the **thyrohyoideus**", "**Rhomboideus major** and **rhomboideus minor** are not distinct muscles", "Tendon of **flexor digitorum profundus** to digit 1 is vestigial or absent", and "**Flexor carpi radialis** originates from the radius"; NB, the genus *Pan* is included in the other Hominini sub-tribe, the Panina).

One of the reversions that occurred within the sub-tribe Hominina, z', violates Dollo's law (i.e., a structure that was phylogenetically lost was later reacquired). Both the rhomboideus major and rhomboideus minor are plesiomorphically present in Euarchontoglires and then, at the node leading to the last common ancestor (LCA) of dermopterans and primates (c.88.8 million years ago (MY): Fig. 6.1), the two muscles became fused, but they appear again as separate muscles in *Homo* (c.2.4 MY: Fig. 6.1). This case is therefore similar to the striking example reported by Wiens (2011) concerning the loss of mandibular teeth in the ancestor of modern frogs and their reappearance in the frog genus *Gastrotheca* (see above). The alternate hypothesis, that the rhomboideus minor was lost independently in all other Primatomorpha taxa, is cladistically very unlikely. Ten evolutionary steps are required for independent losses in dermopterans, strepsirrhines, tarsiers, *Pithecia, Aotus, Saimiri,* hylobatids, orangutans, gorillas and chimpanzees versus only three steps for a reversion. The three steps are unambiguously optimized in the tree reflect loss in Primatomorpha and reacquisition in *Homo* and *Callithrix*. *Callithrix* and the Cercopithecinae usually have a rhomboideus major and minor, but it is not clear whether the Colobinae have both these muscles. Therefore, the rhomboideus minor may have become a distinct muscle either in the subfamily Cercopithecinae or in the family Cercopithecidae (which includes both the Cercopithecinae and the Colobinae).

Eight other reversions within the tree shown in Fig. 6.1 also violate Dollo's law. Some of the alternate hypotheses (i.e., no violation of Dollo's law) for most of these cases are just as unlikely as the alternate hypothesis in the previous paragraph (see Diogo and Wood, 2012b, for more details). These eight cases include: a') Reversion of **"Biceps brachii** has no bicipital aponeurosis" (apart from *Lemur* and *Loris* within the euarchontan taxa examined; the aponeurosis is also usually present in hominoids except *Pongo*, so it is not clear whether hominoids reacquired the aponeurosis and it was subsequently lost again in *Pongo* or it was reacquired independently in the Hylobatidae and the Homininae); c') Reversion of **"Spinotrapezius** is not a distinct muscle"; i') Reversion of "Rhomboideus major and rhomboideus minor are not distinct muscles" in *Callithrix* (see above); m') Reversion of **"Sphincter colli profundus** is not a distinct muscle" (either the muscle was lost in anthropoids and then reappeared in the Cebidae+Aotidae clade and in *Cercopithecus*, or it was lost in *Pithecia* and catarrhines and then reappeared in *Cercopithecus*); r') Reversion of **"Pterygopharyngeus** is not a distinct muscle" (either the derived condition was acquired in Euarchonta and then reverted in *Cynocephalus* and *Hylobates*, or it was acquired in *Tupaia* and Primates and then reverted in *Hylobates*); v') Reversion of **"Epitrochleoanconeus** is not a distinct muscle"; w') Reversion of **"Contrahentes digitorum** are missing"; and x') Reversion of **"Flexores breves profundi** are fused with the **intermetacarpales**, forming the **interossei dorsales**" (either the derived condition was acquired in *Tupaia*, *Cynocephalus*, platyrrhines and hominoids and then reverted in *Pan*, or it was acquired in Euarchonta and then reverted in strepsirrhines, *Tarsius*, cercopithecids, and *Pan*).

6.3 Atavisms, birth defects, "recapitulation", adaptive plasticity, and developmental constraints

The development of the muscular system is a complex process that involves, on the one hand, the fusion and/or reabsorption of elements and, on the other hand, the splitting and/or the neomorphic origin of elements (see, e.g., Diogo and Abdala, 2010). One might think that this system would be particularly prone to evolutionary reversions leading to either decrease or increase in the total number of muscles in the descendant taxa. In our phylogenetic optimization using soft tissue characters, the average time between a structure's loss and reacquisition was *c.*50.9 MY (Diogo and Wood, 2012b). In the nine cases summarized by Wiens (2011; his table 2 plus his *Gastrotheca* example), the mean time between loss and reacquisition was *c.*72.8 MY ([(25 + 32)/2, for the reappearance of a larval stage in plethodontid salamanders] + [40, for the reappearance of a larval stage in hemiphractid frogs + [(95 + 120)/2, for the reappearance of digits in amphisbaeninan lizards of the genus *Bipes*] + [(20 + 60)/2, for the reappearance of digits in gymnophthalmid lizards of the genus *Tretioscincus*] + [15, for the reappearance of digits in gymnophthalmid lizards of the genus *Bachia*] + [(20 + 35)/2, for the reappearance of digits in scincid lizards of the genus *Scelotes*] + [(30 + 80)/2, for the reappearance of eggshell in the boid snake *Eryx*] + [(41 + 79)/2, for the reappearance of shell coiling in snails] + [(225 + 338)/2, for the reappearance of mandibular teeth in the frog genus *Gastrotheca*]/9). However, if the outlying *Gastrotheca* example is excluded, the mean would be similar to that obtained in our study (*c*.46.7 MY), suggesting that muscles are not more prone to evolutionary reversions than are external or hard tissue structures. Along the same lines, results of recent myology-based cladistic analyses of the higher-level phylogeny of teleostean and of bony fish revealed that muscles are generally less prone to homoplasy than are skeletal structures (for a review, see Diogo and Abdala, 2010). Wiens (2011) recognized that his results clearly contradict Dollo's law, and results reported by Diogo and Wood (2012b) do as well. According to Marshall et al. (1994), structures lost after more than 10 MY can almost never be regained because genes and developmental pathways that are not maintained by selection will decay due to mutational changes. However, in all 18 case studies reported by us and surveyed by Wiens (2011), more than 10 MY had elapsed between a structure's loss and its reacquisition.

A flaw in Marshall et al.'s (1994) explanation is that the phylogenetic loss of structures does not always involve a complete loss of the developmental pathways. As noted by Wiens (2011), recent studies suggest that most developmental pathways for tooth development are maintained in at least some birds (e.g., chicken), despite the absence of teeth in adult birds over the last 60 MY. Our comparative analyses strongly support the idea that such reacquisition of anatomical structures often occurs in cases where the developmental pathways were actually maintained (N.B., many developmental pathways participate in the formation of multiple anatomical structures and, therefore, may be maintained by selection, e.g., Gould, 1977; 2002; see below). An illustrative example concerns the presence/absence of the **contrahentes digitorum** in adult hominids. As explained above, chimpanzees display reversion of a synapomorphy of Hominidae (acquired at least 15.4 MY ago: Fig. 6.1) in which adult individuals have two contrahentes digitorum other than the **adductor pollicis** (other adult hominids usually have none), one going to digit 4 and the other to digit 5. According to Marshall's (1994) theory, the genes and developmental pathways needed to form the contrahentes should no longer be present after 15.4 MY of evolution. However, detailed studies of the development of the hand muscles (e.g., Cihak, 1972) have shown that karyotypically normal modern human embryos *do have* contrahentes going to various fingers and that these muscles are lost during later embryonic development (Fig. 6.2A). Moreover, other studies (e.g., Dunlap et al., 1986) have shown that, in karyotypically abnormal modern humans such as individuals with Trisomies 13, 18, or 21, the contrahentes often persist until well after birth (Fig. 6.2B). Our dissections of individuals with these trisomies have confirmed that these muscles are often present in such individuals (Fig. 1.4). Therefore, the presence of the contrahentes in adult chimpanzees probably results from heterochrony (specifically, paedomorphism) in the lineage leading to the genus *Pan*. Similarly, the evolutionary reversion leading to the presence of distinct intermetacarpales in adult chimpanzees (Fig. 6.1) is probably also results from paedomorphism, because the **intermetacarpales** *are present* as distinct muscles in early embryos of karyotypically normal modern humans (Cihak, 1972). Therefore, we use the terms "evolutionary reversion" and "atavism" only to refer to non-embryonic stages; for example: in human individuals with trisomy, *evolutionary reversions* often lead to *atavistic* contrahentes in later stages of development, including adulthood. Since their evolutionary origins, these muscles have always been present in the embryos of the direct ancestors of our species, *H. sapiens*, and continue to be present normally in the embryos of our species. Although these terms can be correctly used in an evolutionary context to refer to the presence of these muscles in non-embryonic stages of modern humans, the muscles never completely disappeared from the earlier ontogenetic stages within the evolutionary history of our lineage.

According to some authors, complex structures that form early in ontogeny just to become lost or indistinct in later developmental stages (so called 'hidden variation') may increase ontogenetic potential early in development. If there are external perturbations (e.g., climate change, environment occupied by new species, etc.), selection can act on that potential (adaptive plasticity) (e.g., West-Eberhard, 2003). However, authors such as Gould (1977) and Alberch (1989) argue that examples such as those cited above support a "constrained" rather than an "adaptationist" view of evolution. This interpretation is also favored by recent authors such as Galis and Metz (2007: 415-416), who stated, "without denying the evolutionary importance of phenotypic plasticity and genetic assimilation, we think that for the generation of macro-evolutionary novelties the evidence for the impact of hidden variation is, thus, far limited." We are inclined to agree that hidden variation has limited potential to generate evolutionary novelties, but it may play a major role in the *reappearance* of some traits associated with these novelties, such as anatomical reversions that violate Dollo's law.

In *Ontogeny and Phylogeny,* Gould argues that, although Haeckel's hypothesis that the ontogeny of one organism recapitulates the adult stages of its ancestors (i.e., recapitulation) has been refuted, researchers often use this idea as a "straw-man" to deny that there is often a parallel between ontogeny and phylogeny. According to Gould, such a parallel exists and is probably driven more by phylogenetic/ontogenetic constraints than by adaptive plasticity. For example, the contrahentes and intermatacarpales appear early on and then become lost or indistinct later in "normal" modern human ontogeny; during the recent evolutionary history of modern humans, these muscles were plesiomorphically present and then were lost. This is not an example of recapitulation in the Haeckelian sense: the contrahentes digitorum and the intermetacarpales of karyotypically 'normal' human embryos do not correspond to the muscles of *adult* primates, such as chimpanzees, or of other primate/mammalian adults, but instead to the muscles of the *embryos* of these taxa. Even after several millions of years, the developmental pathways to produce these muscles in adults have not been completely lost, probably because of ontogenetic constraints (i.e., these pathways also play a role in the formation of other structures that *are* present and functional in modern human adults). Evolutionary reversions that resulted in the presence of the contrahentes and distinct intermetacarpales in extant adult chimpanzees might be either the result of adaptive evolution or a by-product of paedomorphic events related to other structural adaptations. It seems unlikely that the persistence of contrahentes in the later ontogenetic stages of karyotypically abnormal modern humans, such as individuals with Trisomies 21, and particularly 13 and 18 (who do not usually survive long after birth), is the result of adaptive evolution and natural selection. The presence of distinct contrahentes digitorum and intermetacarpales in adult chimpanzees is very likely due to prolonged or delayed development of the hand musculature in these apes; in this respect, extant chimpanzees are more neotenic than modern humans. Although it is often stated that modern humans are more neotenic than other primates, both paedomorphic and peramorphic processes have been involved in the mosaic evolution of humans and of other hominoids (see, e.g., Bufill, 2011, and references therein).

6.4 Future directions: Down syndrome, muscle dysfunction, mouse models, genetics, and apoptosis

Together with our colleague Randall Roper, we are expanding our study of muscle birth defects toward experimental and mechanistic developmental and genetic studies, including the use of mouse models for Down syndrome. As explained above, Down syndrome (DS) results from trisomy of human chromosome 21 (Hsa21); it occurs in approximately 1 of 800 live births and exhibits nearly 80 clinically defined phenotypes with varying severity (e.g., Epstein, 2001; Nadel, 2003; Van Cleve et al., 2006; Van Cleve and Cohen, 2006; Cleves et al., 2007). Previous studies have reported the occurrence of numerous abnormalities of the skeletal musculature, including the frequent presence of additional muscles, such as the hand muscles (contrahentes) and the facial muscle (**platysma cervicale**), in human fetuses, neonates, and adults with DS (e.g., Aziz, 1980; 1981a,b; Dunlap et al., 1986). There is evidence that persons with DS have very low skeletal muscle strength, and hypotonia (low muscle tone) is present in almost all babies with DS (e.g., Korenberg et al., 1994; Cowley et al., 2012). Craniofacial and skeletal deficiencies related to DS may be exacerbated by hypotonia, growth retardation, and low muscle strength (e.g., Hawli et al., 2009). Skeletal muscle defects and reduction in physical function limit ability for independent living, vocational opportunity and productivity, and economic self-sufficiency in this population, often leading to assisted living and lower quality of life; moreover, mobility impairments are predictive of mortality in adults with DS (Cowley et al., 2012).

A large body of literature has documented delays in basic motor skills such as walking, reaching and grasping, motor impairments, and abnormalities in postural and gait control, and movements are generally slower and more variable in children with DS (Carvalho and Vasconcelos, 2011). The causes of these conditions, as well as low strength and other dysfunctions of skeletal muscles in people with DS, are debated; possible explanations include cognitive limitations, neurological disorder, abnormal sensorimotor integration, compromised somatosensory system, and biomechanical deficits (Carvalho and Vasconcelos, 2011). This debate is hampered by an historical lack of focus on the development and phenotype of skeletal muscles and the phenotypic and/or histological causes contributing to muscle dysfunction. For instance, although the Ts65Dn mouse model of DS displays reduced grip strength, running speed, motor coordination, and swimming speed (e.g., Costa et al., 1999), this model has rarely been used to gain insight into the skeletal muscle abnormalities and dysfunction that are so prevalent in humans with DS (Cowley et al., 2012).

The main aim of our current research is to use mouse models of DS to study the specific developmental mechanisms, including cellular (apoptotic study) and genetic (study of Ts65Dn x *Rcan1+/*), related to atypical development, abnormal phenotype, and dysfunction of skeletal muscles in humans with DS. Our main hypothesis is that at least some of these conditions—in particular, the frequent persistence of additional muscles such as the contrahentes and platysma cervicale—are associated with developmental delay; specifically, with decrease in muscle apoptosis due to the presence of an extra copy of the gene *Rcan1*. This project will lead to a better general understanding of DS and other skeletal muscle dysfunctions, and also, potentially, of hypotonia, which is one of the most common symptoms of DS. By focusing on the less-studied skeletal muscles, we aim to also provide a more complete understanding of the entire muscular system in people with DS, including the heart malformations that are the principal cause of mortality in the first two years of life (Stoll et al., 1998) and are very likely related to a decrease of cardiac muscle apoptosis (e.g., Gotlieb, 2009).

The study of muscle development in DS has great potential to improve human health. For instance, it has been argued that humans with DS show increased apoptosis in cells such as neurons, granulocytes, peripheral blood cells, and lymphocytes (e.g., Elsayed and Elsayed, 2009). If our hypothesis of decreased apoptosis in cardiac and skeletal muscles in these individuals is supported, it would reveal mosaicism in apoptosis, thus pointing out the danger of trying to correct some of the conditions found in DS by aiming to decrease apoptosis in a non-targeted way. A possible mismatch between the nervous and muscular systems (i.e., more apoptosis in neurons and less apoptosis in muscle cells and presence of extra muscles) could explain why hypotonia is present in almost all babies with DS. By correlating these data with data on heart malformations in mouse models for DS, specifically exploring the possible association between the *Rcan1* trisomic gene found in humans with DS and in mouse models (such as Ts65Dn) and decrease in both skeletal and cardiac muscle apoptosis, we may better understand muscular abnormalities and dysfunctions and thus facilitate treatment of humans with DS.

The DS genotype-phenotype correlation is an intriguing topic for investigation using mouse models for DS (e.g., Escorihuela et al., 1995; Siarey et al., 1997; Baxter et al., 2000; Olson et al., 2004; 2007; Lorenzi and Reeves, 2006; Alderidge et al., 2007; Belichenko et al., 2009). Using these models, we can analyze embryos and tissues at any stage and observe the mechanisms that produce phenotypes common in DS. Randall Roper is a pioneer in developing and studying these models, in particular the most widely used strain, Ts65Dn (B6EiC3Sn a/A-Ts(17^{16})65Dn/J). This strain of mouse has three copies of orthologs for about half of the genes on Hsa21, including *Rcan1*, and exhibits DS phenotypes including cognitive, craniofacial, cardiac and skeletal deficits (Reeves et al., 1995; Richtsmeier et al., 2000; Moore, 2006; Blazek et al., 2011). A different strain, Tc1 (B6129S-Tc(Hsa21)1TybEmcf/J), carries an extra copy of most (81%) of Hsa21, but not *Rcan1*, and has been used to study cognitive, craniofacial, cardiac and tumor phenotypes associated with DS (O'Doherty et al., 2005; Morice et al., 2008; Dunlevy et al., 2010; Reynolds et al., 2010). Ts1Rhr (B6.129S6-Dp (16Cbr1-ORF9)1Rhr/J) has 33 trisomic genes in 3 copies (about 1/3 of the trisomic genes in Ts65Dn, but only two copies of *Rcan1*) and has been used to investigate the contribution of these genes to cognitive, cardiac, craniofacial, tumor and skeletal phenotypes (Olson et al., 2004; Belichenko et al., 2009; Dunlevy et al., 2010; Olson and Mohan, 2011). Ts1Yey (B6;129S7-Dp(16Lipi-Zfp295)1Yey/J) has a triplication of the entire region of Hsa21 that is homologous to mouse chromosome 16, including *Rcan1,* and exhibits cognitive and cardiac phenotypes associated with DS (Li et al., 2007; Yu et al., 2010; Liu et al., 2011). Ts65Dn and Tc1 mice are maintained on a ~50% B6, C3H or ~50% B6, 129S background and cannot be inbred. Ts1Rhr and Ts1Yey mice have been backcrossed to B6 mice for multiple generations.

Recent molecular and developmental studies using mouse models for DS provide indirect evidence to support decrease in apoptosis as a cause of the frequent persistence of additional muscles until later stages of development in trisomic individuals. A small family of proteins, termed MCIP1 and MCIP2 (myocyte-enriched calcineurin interacting protein), are expressed most abundantly in skeletal and cardiac muscles and form a physical complex with calcineurin A. MCIP1 is encoded by *Rcan1*, a trisomic gene found in Ts65Dn and Ts1Yey mice as well as humans with DS (e.g., Rothermel, 2000; Gotlieb, 2009). Expression of the MCIP family of proteins is up-regulated during muscle differentiation, and its forced over-expression inhibits calcineurin signaling leading to a decrease in muscle apoptosis. A recent study of the Ts65Dn mouse model for DS showed that, when apoptosis is abnormally reduced in trisomic mice embryos, excess populations of myocytes may form in the atrioventricular region. These myocytes may interfere with the normal migration of cells during cardiac development, leading to valvular abnormalities and atrioventricular or ventricular septal defects similar to the congenital heart defects typical of humans with DS (Gotlieb, 2009). These mice also seem to demonstrate developmental delay (e.g., Blazek et al., 2010). However, there are no detailed studies on the morphology, development, or apoptosis levels of skeletal muscles in any mouse model for DS.

From our previous studies on the evolution, homologies and development of the skeletal muscles of all major vertebrate taxa (e.g., Diogo, 2007; Diogo et al., 2008a,b; 2009a,b;

2012a,b; 2013; Diogo and Abdala, 2010; Diogo and Wood, 2011; 2012a,b; Diogo and Tanaka, 2012) we are able to: 1) identify homologies between mouse and human muscles and judge whether muscle abnormalities in mouse models for DS mirror those found in humans with DS; and 2) compare the normal ontogeny of mouse muscles with each developmental stage of the Ts65Dn trisomic specimens and determine which muscles show atypical development. By combining these data with the data in this book on muscle variations found in karyotypically normal humans and abnormalities found humans with DS, we can estimate percentages of muscle abnormalities in humans with DS in comparison to each of the mouse strains. The frequent presence of skeletal muscles in humans with DS that were "lost" in evolution, likely due to developmental delay (as discussed in the previous section), may be related to a decrease in muscle apoptosis.

Roper and colleagues have performed numerous studies on mouse models for DS (e.g., Roper et al., 2006a,b; 2009; Roper and Reeves, 2006; Blazek et al., 2010; 2011; Solzak et al., 2013; Billingsley et al., in press), including interdisciplinary investigations of craniofacial and limb skeletal phenotypes. They provided the first experimental evidence that trisomy affects neural crest cells (NCC): quantification in Ts65Dn E9.5 mice found deficits in trisomic cranial NCC generation, migration, and proliferation affecting PA1 (Roper et al., 2009). Other experiments using rhombomeric quail-to-chick grafts indicated that rhombomeric populations remain coherent during ontogeny, with rhombomere-specific matching of muscle connective tissue with its attachment sites for branchial and tongue muscles (Köntges and Lumsden, 1996). These results emphasize the importance of NCC for the typical patterning of muscles and help to explain how changes in the NCC in DS could affect the configuration and attachments of head muscles. Increased apoptosis during development has been observed in the hippocampus of humans with DS, but Roper and colleagues found no differences in apoptotic cells in neuronal precursors in Ts65Dn mice (Chakrabarti et al., 2007; Guidi et al., 2008). The authors examined trisomic and euploid E10.5 mice using an antibody for cleaved caspase 3 (c-caspase 3), revealing increased apoptosis in trisomic (Ts65Dn) embryos in craniofacial and neurological precursors (Fig. 1; Solzak et al., 2013).

Recently, Roper and the authors performed a preliminary analysis of the phenotype of skeletal muscles of Ts65Dn mice. We dissected two heads and four forelimbs of 0-day old (P0, neonate) mice, two heads and four forelimbs of six week old (6W) mice, one head and two forelimbs of one year old mice, and four forelimbs of adult trisomic mice and compared these mice with euploids from the Ts65Dn strain. At least some trisomic mice have additional muscles that are usually not present in euploids, especially in the pectoral region. The topological position of these pectoral muscles is similar to that of additional pectoral muscles that are often present in humans with DS, namely the **pectoralis abdominalis**, **pectorodorsalis** and **chondrohumeralis** (e.g., Dunlap et al., 1986; see Chapters 2 and 3). In the near future, we plan to dissect the head, forelimb, and hindlimb muscles of several trisomic and euploid mice and compare them with the mice we have already dissected. Our aims are to: 1) rigorously quantify muscle abnormalities in Ts65Dn mice; 2) test the hypothesis that muscle weights are higher in trisomic mice than euploid mice, presumably due to decrease in muscle apoptosis; 3) test the hypothesis that at least some muscle abnormalities of trisomic mice are related to developmental delay; and 4) compare frequencies of muscle abnormalities between the forelimb and hindlimb (in aneuploid humans, forelimb muscle abnormalities are more frequent than hindlimb ones; Barash et al., 1970). Jones (1979) showed that the hindlimb muscles **tenuissimus** and **sartorius** appear early in the ontogeny of karyotypically normal mice and then usually disappear during embryonic development. In this respect, the ontogeny of these muscles is similar to that of the **contrahentes digitorum** and **platysma cervicale** of karyotypically normal humans, which normally appear and then disappear at early stages but are present at older stages in humans with DS. Therefore, presence of the tenuissimus/sartorius in P0 or 6W Ts65Dn trisomic mice would strongly support our hypothesis of delayed muscle development. We also plan to dissect other mouse models of Down syndrome and perform gene expression studies of *Rcan1* and apoptotic studies in all these strains; by doing so, we hope to shed light on potential genotype-phenotype correlations in mouse models and gather further evidence to test our hypothesis that at least some muscle abnormalities of trisomic mice and humans are related to developmental delay associated with decrease in muscle apoptosis.

In summary, the study of normal and abnormal muscle development has crucial implications for macroevolution and human evolution in particular, development, and birth defects. We aim to integrate current understanding of DS in humans with comparative studies of vertebrate evolution and mouse models with a focus on the phenotypic, cellular, and genetic causes of atypical development. In addressing these central questions from a broad, interdisciplinary comparative and evolutionary perspective, we hope to pave the way for further mechanistic analyses of the regulators associated with the control of muscle development in DS, and ultimately lead to a better understanding of human trisomies and facilitate testing of potential treatments for musculoskeletal phenotypes commonly found in DS.

Illustrations

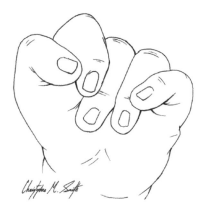

Figure 1.1. Overlapping digits of hand characteristic of Trisomy 18.

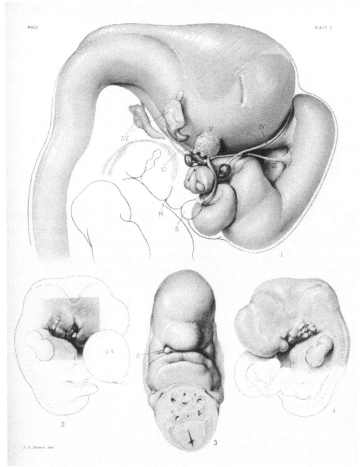

Figure 1.2. Human cyclopic embryo. 1. The developing brain and cyclopic eye of the embryo. Cranial nerves are marked with roman numerals. **S:** snout, **M:** mouth, **OV:** optic vesicle. 2. Right side of cyclopic embryo. **UV:** umbilical vesicle. 3. Ventral surface of embryo. S: snout. 4. Left side of embryo. Illustrated by James Didusch (Mall, 1917). Used with permission. Department of Art as Applied to Medicine, Johns Hopkins University School of Medicine.

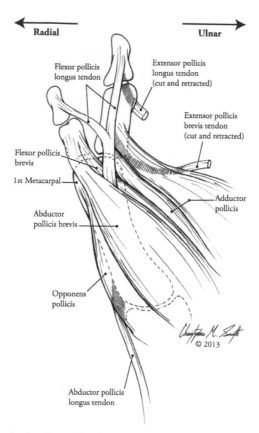

Figure 1.3. Scheme illustrating the hand muscles of the four-fingered hand of a human newborn with Trisomy 18 that represents an extreme case of limb birth defects (the other hand having six digits and being shown in Fig. 1.4). Our hypothesis is supported because, despite the absence of a thumb, all the muscles normally associated with the thumb are present and attach to digit 2, which is the most radial digit. The only exception to our hypothesis is that in this hand the most radial tendon of the flexor digitorum superficialis, which usually goes to digit 2 (i.e., to a digit that is not the most radial digit) goes to digit 2 of this hand (i.e., to its most radial digit).

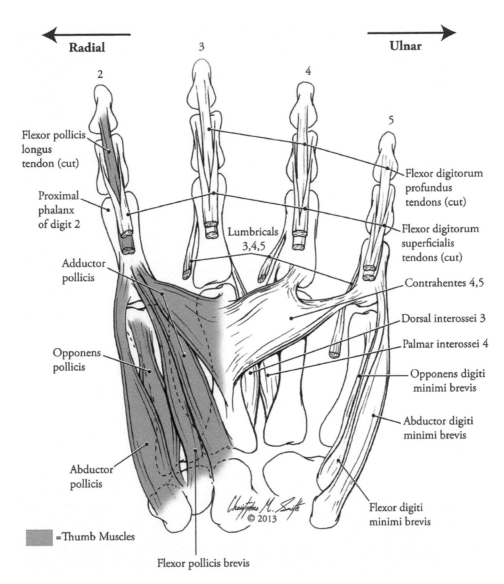

← **Radial**

3

4

2

5

Flexor pollicis longus tendon (cut)

Proximal phalanx of digit 2

Adductor pollicis

Opponens pollicis

Abductor pollicis

Lumbricals 3,4,5

Ulnar →

Flexor digitorum profundus tendons (cut)

Flexor digitorum superficialis tendons (cut)

Contrahentes 4,5

Dorsal interossei 3

Palmar interossei 4

Opponens digiti minimi brevis

Abductor digiti minimi brevis

Flexor digiti minimi brevis

=Thumb Muscles

Flexor pollicis brevis

© 2013

Figure 1.4. Scheme illustrating the thumb muscles of the other hand (the one with six digits) of the same human newborn with Trisomy 18. Our hypothesis is supported because, despite the presence of two thumbs (which are the only digits illustrated in this figure), the muscles normally associated with the thumb are not duplicated. Instead, the muscles that normally insert respectively onto the radial and ulnar sides of the thumb insert onto the radial and ulnar sides of the most radial and most ulnar thumbs, respectively, as predicted. Interestingly, the tendon of the flexor digitorum profundus, which usually goes to the central (so, not ulnar and not radial) portion of the thumb, bifurcates to go to both the most ulnar and most radial thumbs.

Color image of this figure appears in the color plate section at the end of the book.

Normal

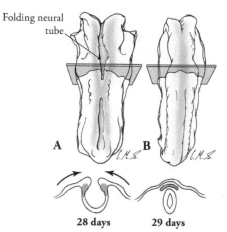

Folding neural tube

A   **B**

28 days   29 days

Neural crest cells originate from the "crest" of the folding neural tube and reside between the neural tube and ectoderm after fusion

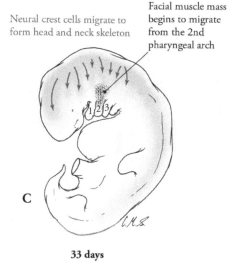

Neural crest cells migrate to form head and neck skeleton

Facial muscle mass begins to migrate from the 2nd pharyngeal arch

C

33 days

Normal   Trisomy 18 Cyclopia

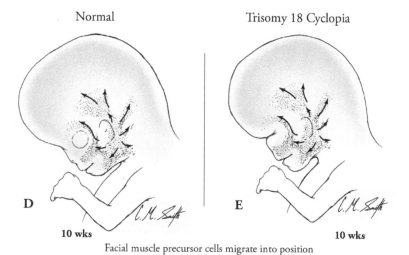

D   **E**

10 wks   10 wks

Facial muscle precursor cells migrate into position across the face attaching onto neural crest cell derived structures and then inserting into skin and cartilage

Figure 1.5. Development of facial musculature. Normal development with neural crest cells depicted in purple, at 28 days **(A)**, 29 days **(B)**, 33 days **(C)**. Comparison of normal **(D)** and Trisomy 18 cyclopic **(E)** facial muscle migration at 10 weeks. Topological position may play a role in facial muscle attachments, as seen from this study (see Chapter 4). Our hypothesis is that in both normal and Trisomy 18 cases, the facial muscles migrate essentially in the same directions, regardless of underlying skeletal structure, then mainly attach to their "nearest neighbor" based on their topological position in space.

Color image of this figure appears in the color plate section at the end of the book.

Right

Left

Normal

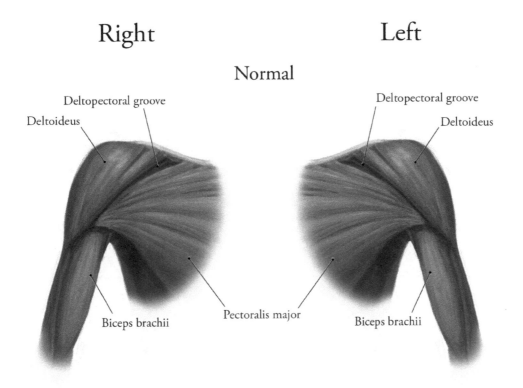

Deltopectoral groove

Deltoideus

Deltopectoral groove

Deltoideus

Biceps brachii

Pectoralis major

Biceps brachii

Trisomy 18 Cyclopia

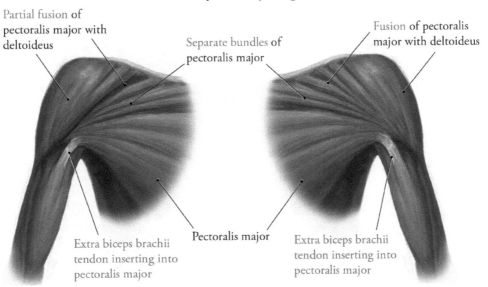

Partial fusion of
pectoralis major with
deltoideus

Separate bundles of
pectoralis major

Fusion of pectoralis
major with deltoideus

Extra biceps brachii
tendon inserting into
pectoralis major

Pectoralis major

Extra biceps brachii
tendon inserting into
pectoralis major

Figure 2.1. Comparison of shoulder musculature in normal and Trisomy 18 cyclopia. Anomalies shown in red.

Color image of this figure appears in the color plate section at the end of the book.

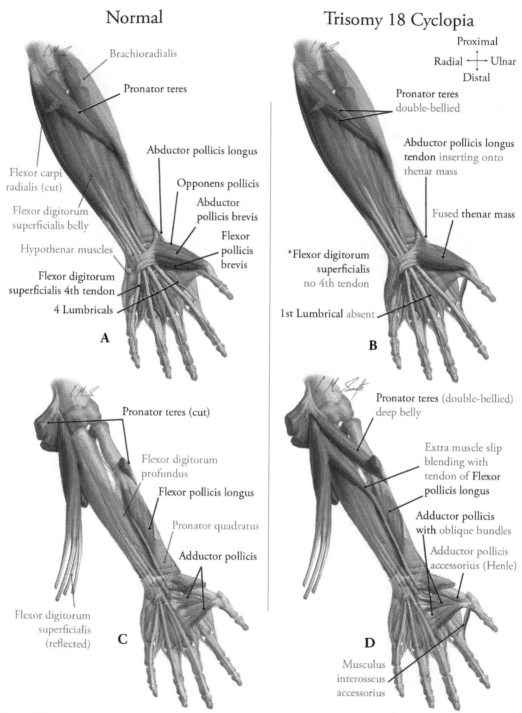

Figure 2.2. Comparison of normal and Trisomy 18 cyclopia left flexors. Superficial muscles **(A, B)**, deep muscles **(C, D)**. Black labels on A and C indicate normal presentation of structures found to be anomalous in Trisomy 18 cyclopia (B, D). Anomalous features (red labels), cartilage (purple), normal structures included for orientation (grey labels).

Color image of this figure appears in the color plate section at the end of the book.

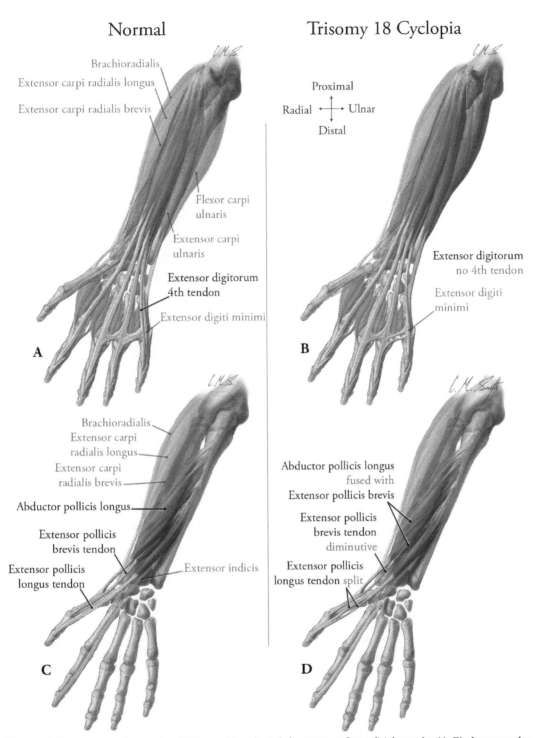

Figure 2.3. Comparison of normal and Trisomy 18 cyclopia left extensors. Superficial muscles **(A, B)**, deep muscles **(C, D)**. Black labels on A and C indicate normal presentation of structures found to be anomalous in Trisomy 18 cyclopia **(B, D)**. Anomalous features (red labels), cartilage (purple), normal structures included for orientation (grey labels).

Color image of this figure appears in the color plate section at the end of the book.

Trisomy 18 Cyclopia Left Forearm and Hand

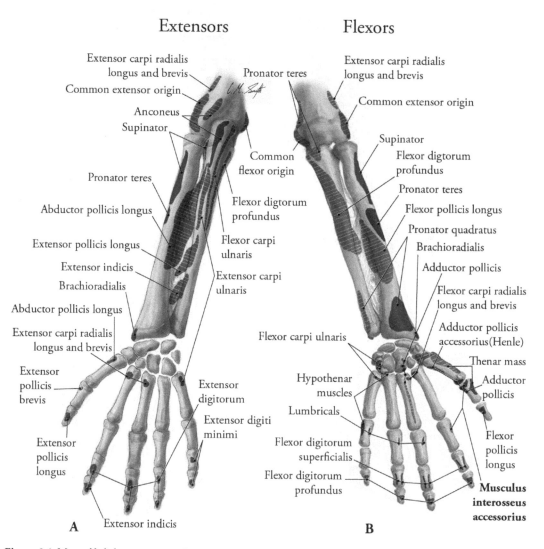

Figure 2.4. Map of left forearm musculature origins and insertions in Trisomy 18 cyclopia. Origins are shown in red (striped). Insertions are shown in blue. Non-ossified cartilage is shown in purple.

Color image of this figure appears in the color plate section at the end of the book.

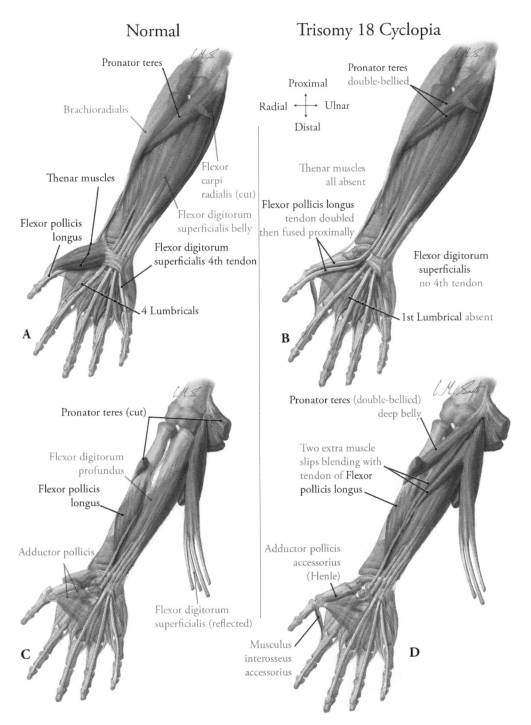

Normal

Trisomy 18 Cyclopia

Pronator teres

Brachioradialis

Thenar muscles

Flexor pollicis
longus

Flexor
carpi
radialis (cut)

Flexor digitorum
superficialis belly

Flexor digitorum
superficialis 4th tendon

4 Lumbricals

A

Proximal

Radial ←→ Ulnar

Distal

Pronator teres
double-bellied

Thenar muscles
all absent

Flexor pollicis longus
tendon doubled
then fused proximally

Flexor digitorum
superficialis
no 4th tendon

1st Lumbrical absent

B

Pronator teres (cut)

Flexor digitorum
profundus

Flexor pollicis
longus

Adductor pollicis

Flexor digitorum
superficialis (reflected)

C

Pronator teres (double-bellied)
deep belly

Two extra muscle
slips blending with
tendon of Flexor
pollicis longus

Adductor pollicis
accessorius
(Henle)

Musculus
interosseus
accessorius

D

Figure 2.5. Comparison of normal and Trisomy 18 cyclopia right flexors. Superficial muscles (**A, B**), deep muscles (**C, D**). Black labels on A and C indicate normal presentation of structures found to be anomalous in Trisomy 18 cyclopia (B, D). Anomalous features (red labels), cartilage (purple), normal structures included for orientation (grey labels).

Color image of this figure appears in the color plate section at the end of the book.

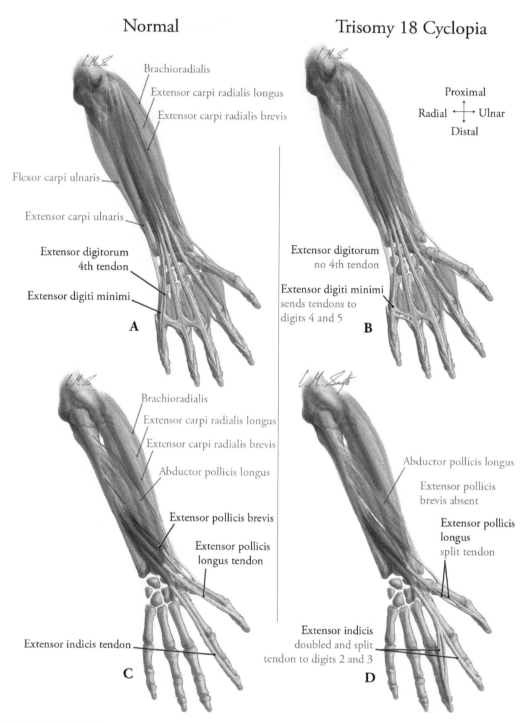

Figure 2.6. Comparison of normal and Trisomy 18 cyclopia right extensors. Superficial muscles **(A,B)**, deep muscles **(C, D)**. Black labels on A and C indicate normal presentation of structures found to be anomalous in Trisomy 18 cyclopia (B, D). Anomalous features (red labels), cartilage (purple), normal structures included for orientation (grey labels).

Color image of this figure appears in the color plate section at the end of the book.

Trisomy 18 Cyclopia Right Forearm and Hand

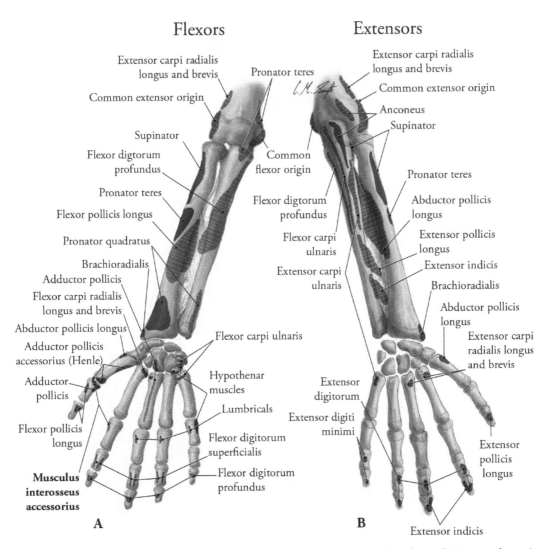

Figure 2.7. Map of right forearm musculature origins and insertions in Trisomy 18 cyclopia. Origins are shown in red (striped). Insertions are shown in blue. Non-ossified cartilage is shown in purple.

Color image of this figure appears in the color plate section at the end of the book.

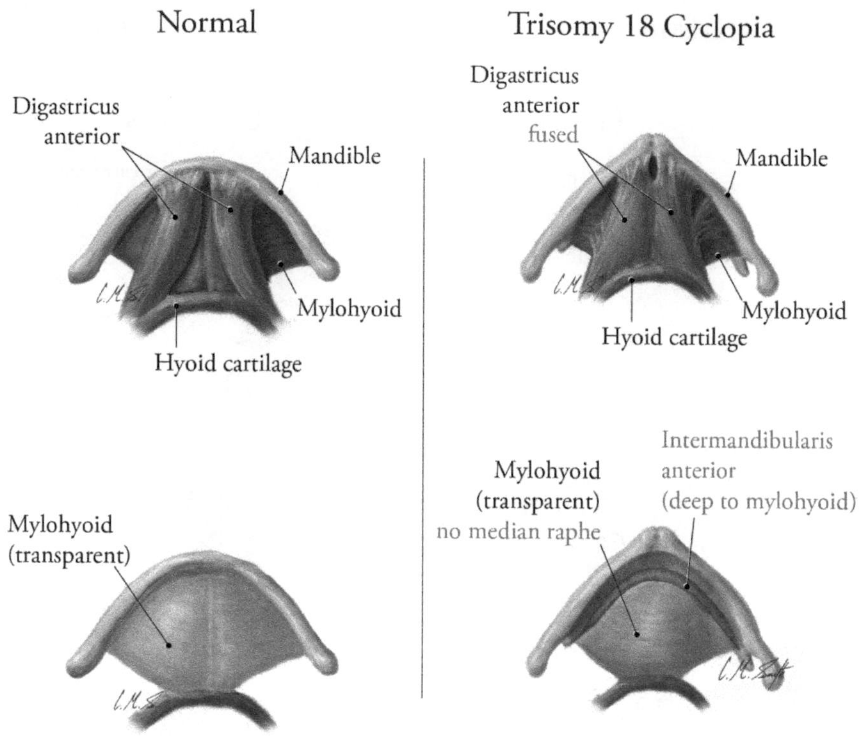

Figure 2.8. Comparison of inferior mandibular view in normal and Trisomy 18 cyclopia. Anomalies labeled in red. Cartilage shown in purple.

Color image of this figure appears in the color plate section at the end of the book.

Normal

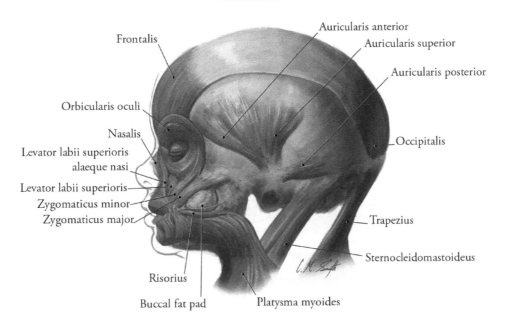

- Frontalis
- Auricularis anterior
- Auricularis superior
- Auricularis posterior
- Orbicularis oculi
- Nasalis
- Levator labii superioris alaeque nasi
- Levator labii superioris
- Zygomaticus minor
- Zygomaticus major
- Occipitalis
- Trapezius
- Sternocleidomastoideus
- Risorius
- Buccal fat pad
- Platysma myoides

Trisomy 18 Cyclopia

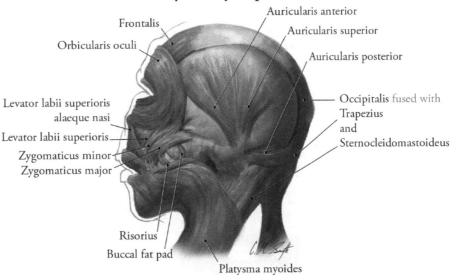

- Frontalis
- Auricularis anterior
- Orbicularis oculi
- Auricularis superior
- Auricularis posterior
- Levator labii superioris alaeque nasi
- Levator labii superioris
- Zygomaticus minor
- Zygomaticus major
- Occipitalis fused with Trapezius and Sternocleidomastoideus
- Risorius
- Buccal fat pad
- Platysma myoides

Figure 2.9. Comparison of superficial lateral head musculature in normal and Trisomy 18 cyclopia. Presence of a platysma cervicale could not determined due to previous dissection.

Color image of this figure appears in the color plate section at the end of the book.

Normal

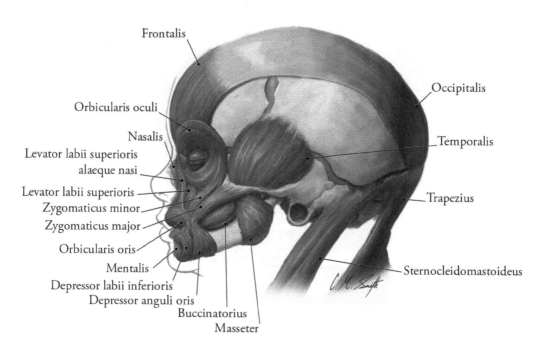

Frontalis

Orbicularis oculi

Nasalis

Levator labii superioris
alaeque nasi

Levator labii superioris

Zygomaticus minor

Zygomaticus major

Orbicularis oris

Mentalis

Depressor labii inferioris

Depressor anguli oris

Buccinatorius

Masseter

Occipitalis

Temporalis

Trapezius

Sternocleidomastoideus

Trisomy 18 Cyclopia

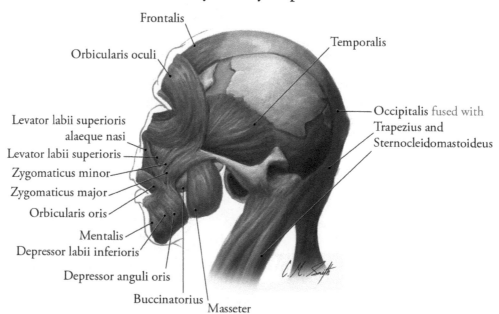

Frontalis

Orbicularis oculi

Temporalis

Levator labii superioris
alaeque nasi

Levator labii superioris

Zygomaticus minor

Zygomaticus major

Orbicularis oris

Mentalis

Depressor labii inferioris

Depressor anguli oris

Buccinatorius

Masseter

Occipitalis *fused with* Trapezius and Sternocleidomastoideus

Figure 2.10. Comparison of lateral head musculature in normal and Trisomy 18 cyclopia. Platysma myoides, risorius, and buccal fat pad removed.

Color image of this figure appears in the color plate section at the end of the book.

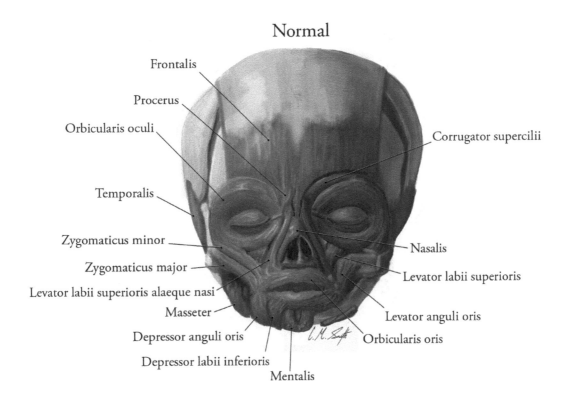

Normal

Frontalis
Procerus
Orbicularis oculi
Corrugator supercilii
Temporalis
Zygomaticus minor
Zygomaticus major
Levator labii superioris alaeque nasi
Masseter
Depressor anguli oris
Depressor labii inferioris
Mentalis
Nasalis
Levator labii superioris
Levator anguli oris
Orbicularis oris

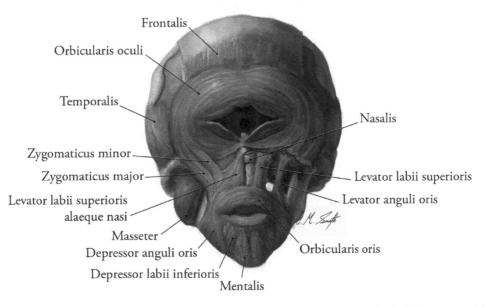

Trisomy 18 Cyclopia

Frontalis
Orbicularis oculi
Temporalis
Zygomaticus minor
Zygomaticus major
Levator labii superioris alaeque nasi
Masseter
Depressor anguli oris
Depressor labii inferioris
Mentalis
Nasalis
Levator labii superioris
Levator anguli oris
Orbicularis oris

Figure 2.11. Comparison of anterior head musculature in normal and Trisomy 18 cyclopia. Platysma myoides, risorius, and buccal fat pad removed. Left side shows deep dissection.

Color image of this figure appears in the color plate section at the end of the book.

Trisomy 18 Cyclopia Lateral Skull

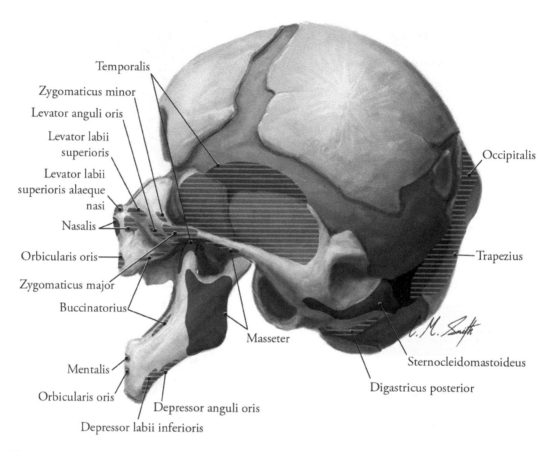

Figure 2.12. Map of lateral skull origins and insertions in Trisomy 18 cyclopia. Origins are shown in red (striped). Insertions are shown in blue. Fontanelles are shown in purple.

Color image of this figure appears in the color plate section at the end of the book.

Trisomy 18 Cyclopia Anterior Skull

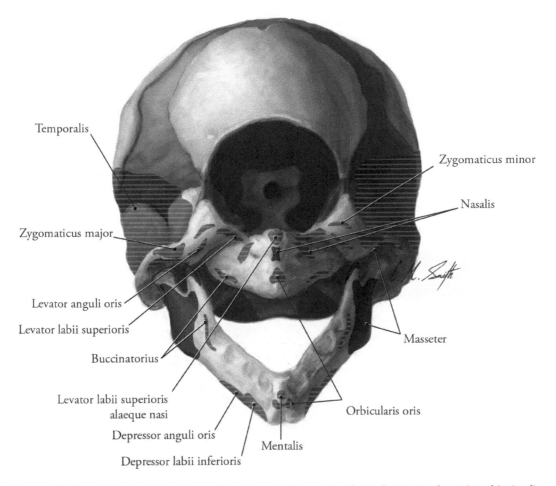

Temporalis

Zygomaticus minor

Zygomaticus major

Nasalis

Levator anguli oris

Levator labii superioris

Buccinatorius

Masseter

Levator labii superioris
alaeque nasi

Depressor anguli oris

Orbicularis oris

Mentalis

Depressor labii inferioris

Figure 2.13. Map of anterior skull origins and insertions in Trisomy 18 cyclopia. Origins are shown in red (striped). Insertions are shown in blue. Fontanelles are shown in purple.

Color image of this figure appears in the color plate section at the end of the book.

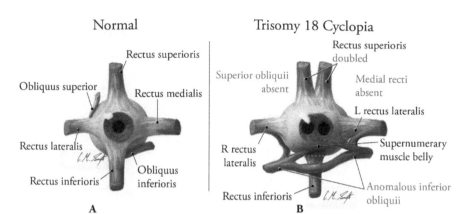

Figure 2.14. Comparison of normal and Trisomy 18 cyclopia eyes. Orbit musculature with cut bellies reflected anteriorly. Normal right eye **(A)**, contrasted with fused irises in Trisomy 18 cyclopia **(B)** Anomalies labeled in red.

Color image of this figure appears in the color plate section at the end of the book.

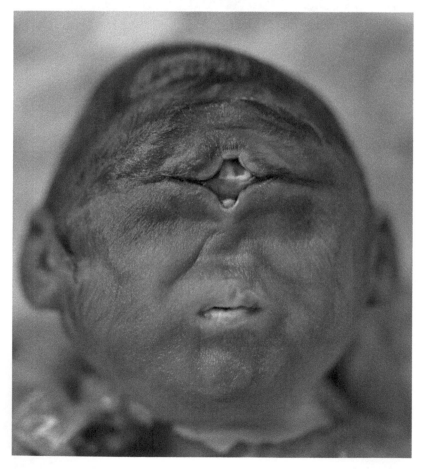

Figure 2.15. Anterior view of the Trisomy 18 cyclopia face. Note the interesting upper eyelid structure of two lids and medial portion.

Color image of this figure appears in the color plate section at the end of the book.

Normal

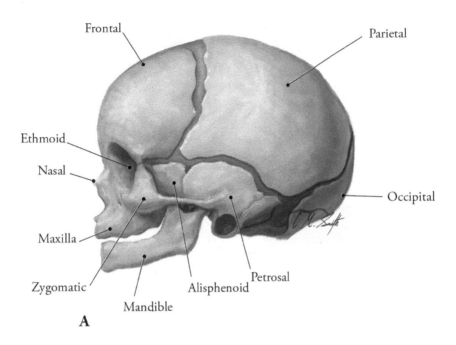

Frontal

Parietal

Ethmoid

Nasal

Occipital

Maxilla

Zygomatic

Petrosal

Alisphenoid

Mandible

A

Trisomy 18 Cyclopia

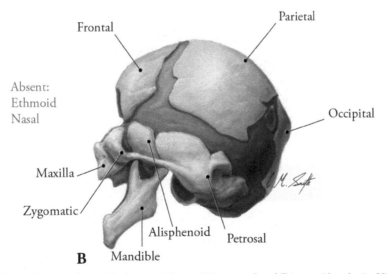

Frontal

Parietal

Absent:
Ethmoid
Nasal

Occipital

Maxilla

Zygomatic

Alisphenoid

Petrosal

B Mandible

Figure 2.16. Lateral comparison of the bones of the skull in normal and Trisomy 18 cyclopia. Visible absent bones are labeled in red.

Color image of this figure appears in the color plate section at the end of the book.

Normal

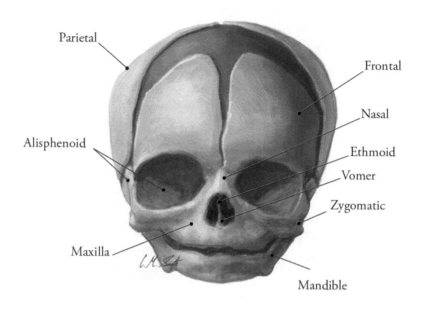

Trisomy 18 Cyclopia

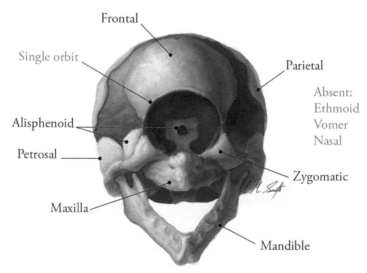

Figure 2.17. Anterior comparison of the bones of the skull in normal and Trisomy 18 cyclopia. Visible absent bones are labeled in red.

Color image of this figure appears in the color plate section at the end of the book.

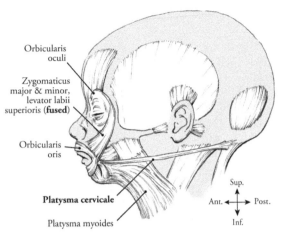

Figure 3.1. Platysma cervicale originating from corner of mouth with no risorius, inserting with the trapezius onto the occipital bone. Fusion of the zygomaticus major, minor and levator labii superioris in a thin sheet originating at the corner of the orbicularis oculi. Anomalies labeled in bold (modified from Bersu, 1980).

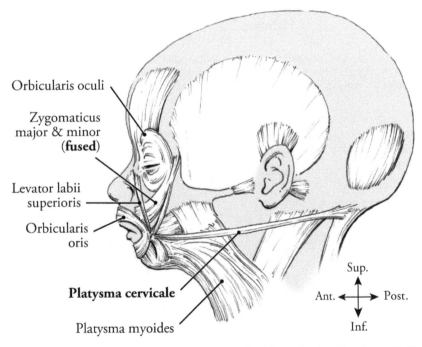

Figure 3.2. Platysma cervicale originating from corner of mouth with no risorius, inserting with the trapezius onto the occipital bone. Fusion of the zygomaticus major and minor originating at the corner of the orbicularis oculi. Anomalies labeled in bold (modified from Bersu, 1980).

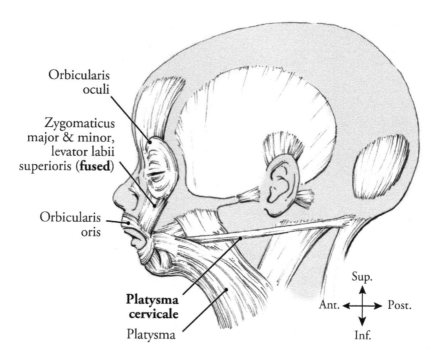

Figure 3.3. Platysma cervicale originating from corner of mouth, inserting with the trapezius onto the occipital bone. Fusion of the zygomaticus major, minor and levator labii superioris in a broad sheet originating from the inferior portion of the orbicularis oculi. Anomalies labeled in bold (modified from Aziz, 1981).

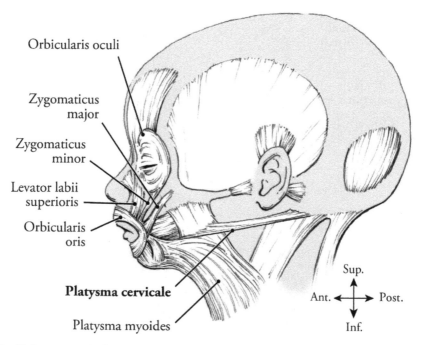

Figure 3.4. Platysma cervicale originating from posterior platysma myoides, inserting with the sternocleidomastoideus onto the mastoid process. Anomalies labeled in bold (modified from Aziz, 1980).

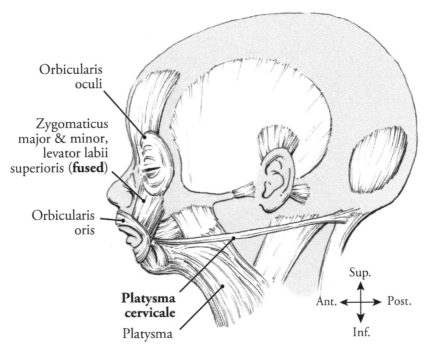

Figure 3.5. Platysma cervicale originating from corner of mouth with no risorius, inserting with the trapezius onto the occipital bone. Fusion of the zygomaticus major, minor and levator labii superioris in a broad sheet originating from the inferior portion of the orbicularis oculi. Anomalies labeled in bold (modified from Aziz, 1980).

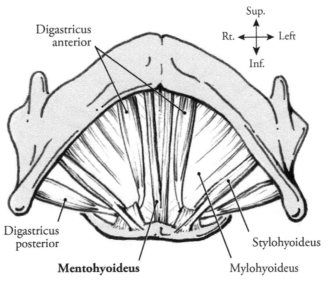

Figure 3.6. Inferior view of hyoid region. Note the presence of the supernumerary mentohyoideus muscle spanning from the medial hyoid to the mandible. Anomalies labeled in bold (modified from Bersu and Ramirez-Castro, 1977).

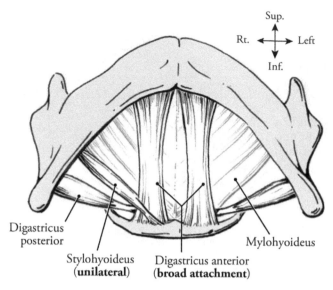

Figure 3.7. Inferior view of hyoid region. Digastricus anterior has broad hyoid attachment with unilateral stylohyoid. Anomalies labeled in bold (modified from Bersu and Ramirez-Castro, 1977).

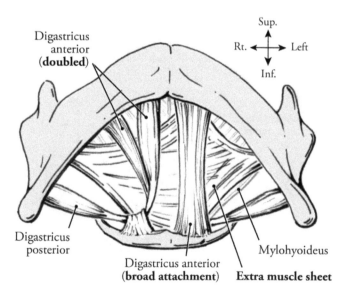

Figure 3.8. Inferior view of hyoid region. Right digastricus anterior doubled. Left digastricus anterior has broad hyoid attachment. Note the extra muscle sheet just superficial to the mylohyoideus. Anomalies labeled in bold (modified from Bersu and Ramirez-Castro, 1977).

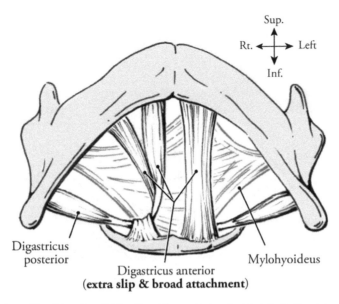

Figure 3.9. Inferior view of hyoid region. Right digastricus anterior doubled. Left digastricus anterior has broad hyoid attachment. Anomalies labeled in bold (modified from Bersu and Ramirez-Castro, 1977).

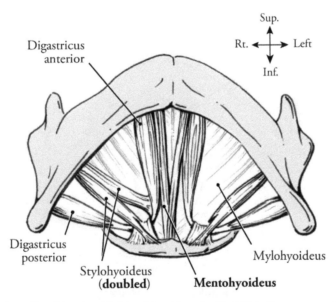

Figure 3.10. Inferior view of hyoid region. Right stylohyoideus doubled. Note the presence of the supernumerary mentohyoideus muscle spanning from the medial hyoid to the mandible. Anomalies labeled in bold (modified from Bersu and Ramirez-Castro, 1977).

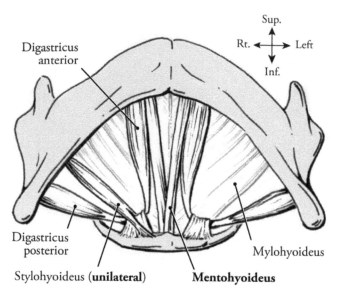

Figure 3.11. Inferior view of hyoid region. Unilateral stylohyoideus. Note the presence of the supernumerary mentohyoideus muscle spanning from the medial hyoid to the mandible. Anomalies labeled in bold (modified from Bersu and Ramirez-Castro, 1977).

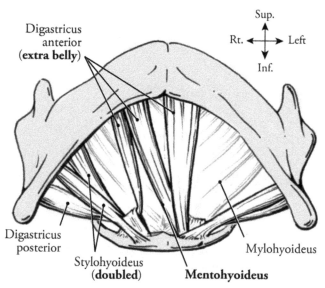

Figure 3.12. Inferior view of hyoid region. Right stylohyiodeus doubled. Note the presence of a single supernumerary mentohyoideus muscle spanning from the medial hyoid to the mandible. Digastricus anterior extra belly fused with mentohyoideus. Anomalies labeled in bold (modified from Bersu and Ramirez-Castro, 1977).

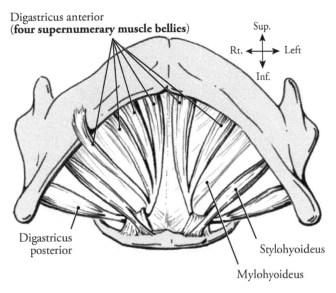

Figure 3.13. Inferior view of hyoid region. Four supernumerary muscle bellies originating from digastricus anterior. Anomalies labeled in bold (modified from Aziz, 1980).

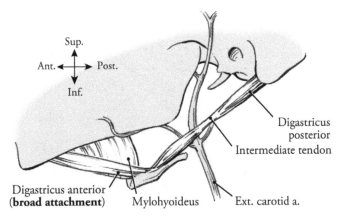

Figure 3.14. Left lateral view of suprahyoid region. Note the high position of the digastricus intermediate tendon. Broad hyoid attachment of left digastricus anterior. Anomalies labeled in bold (modified from Bersu and Ramirez-Castro, 1977).

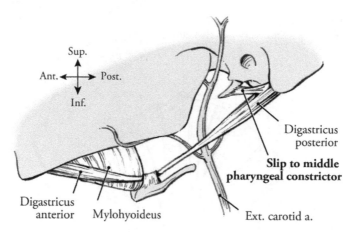

Figure 3.15. Left lateral view of suprahyoid region. The left digastricus posterior gives off a slip to the middle pharyngeal constrictor. Anomalies labeled in bold (modified from Bersu and Ramirez-Castro, 1977).

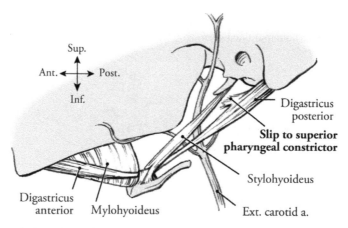

Figure 3.16. Left lateral view of suprahyoid region. The left digastricus posterior gives off a slip to the superior pharyngeal constrictor. Anomalies labeled in bold (modified from Bersu and Ramirez-Castro, 1977).

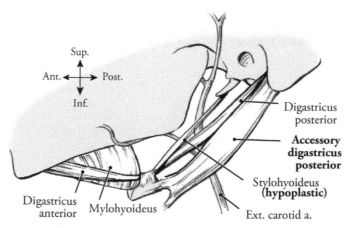

Figure 3.17. Left lateral view of suprahyoid region. Note the presence of the supernumerary accessory digastricus posterior muscle. The stylohyoideus is hypoplastic. Anomalies labeled in bold (modified from Pettersen, 1979).

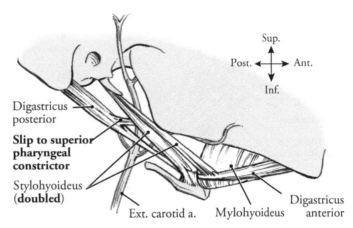

Figure 3.18. Right lateral view of suprahyoid region. The right digastricus posterior gives off a slip to the superior pharyngeal constrictor. Stylohyoideus is doubled. Anomalies labeled in bold (modified from Bersu and Ramirez-Castro, 1977).

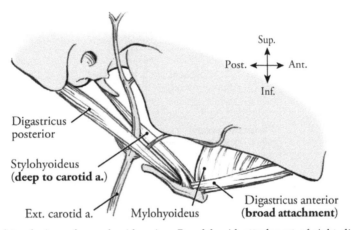

Figure 3.19. Right lateral view of suprahyoid region. Broad hyoid attachment of right digastricus anterior. Anomalies labeled in bold (modified from Bersu and Ramirez-Castro, 1977).

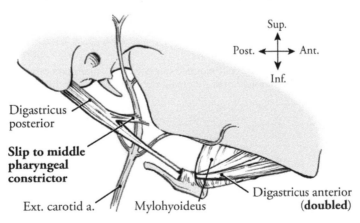

Figure 3.20. Right lateral view of suprahyoid region. The right digastricus posterior gives off a slip to the middle pharyngeal constrictor. Digastricus anterior is doubled. Anomalies labeled in bold (modified from Bersu and Ramirez-Castro, 1977).

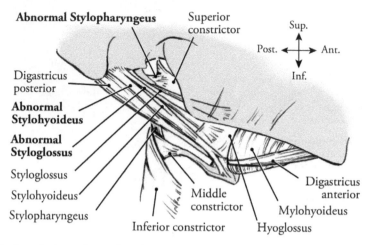

Figure 3.21. Right lateral view of suprahyoid region. Note the abnormal attachments of the stylohyoideus, styloglossus, and the stylopharyngeus. Anomalies labeled in bold (modified from Bersu and Ramirez-Castro, 1977).

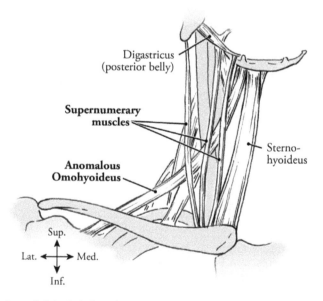

Figure 3.22. Anterior view of right infrahyoid region. Note the three supernumerary muscle bellies and the missing intermediate tendon in the omohyoideus. Anomalies labeled in bold (modified from Aziz, 1979).

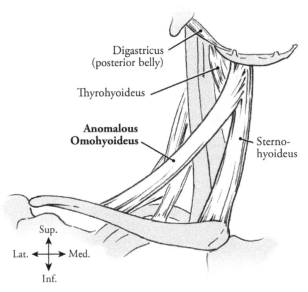

Figure 3.23. Anterior view of right infrahyoid region. Omohyoideus is missing intermediate tendon and partially inserts onto fibers of the sternohyoideus. Anomalies labeled in bold (modified from Aziz, 1980).

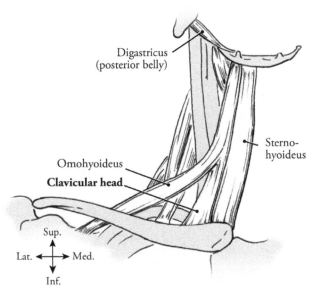

Figure 3.24. Anterior view of right infrahyoid region. Omohyoideus is missing intermediate tendon and gives off an extra clavicular head. Anomalies labeled in bold (modified from Bersu and Ramirez-Castro, 1977).

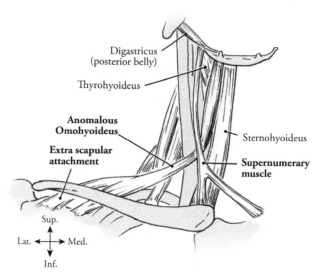

Figure 3.25. Anterior view of right infrahyoid region. Omohyoideus gives off a scapular head. Note the supernumerary muscle attaching from the greater horn of the hyoid to the clavicle and manubrium. Anomalies labeled in bold (modified from Bersu and Ramirez-Castro, 1977).

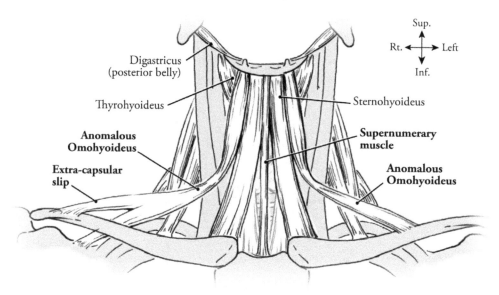

Figure 3.26. Anterior view of infrahyoid region. Omohyoideus gives off an extra-capuslar slip on the right and extra clavicular head on the left. Note the supernumerary muscle just medial to the sternohyoideus muscles. Anomalies labeled in bold (modified from Bersu and Ramirez-Castro, 1977).

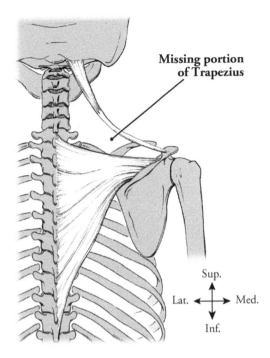

Figure 3.27. Posterior view of the back in human triploidy showing the middle absent portion of the trapezius. Anomalies labeled in bold (modified from Moen, 1984).

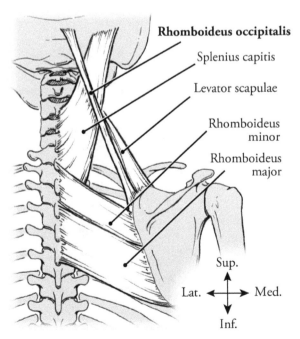

Figure 3.28. Posterior view of the back. The rhomboideus occipitalis spans from the superior angle of the scapula between the rhomboideus minor and levator scapulae to the occipital region of the skull. Anomalies labeled in bold (modified from Aziz, 1981).

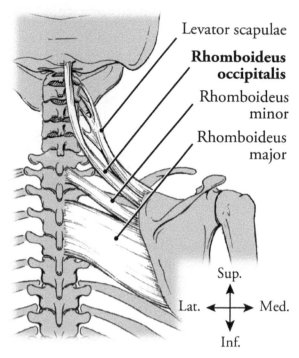

Figure 3.29. Posterior view of the back. The rhomboideus occipitalis spans from the superior angle of the scapula between the diminutive rhomboideus minor and levator scapulae to the occipital region of the skull. Anomalies labeled in bold (modified from Aziz, 1979).

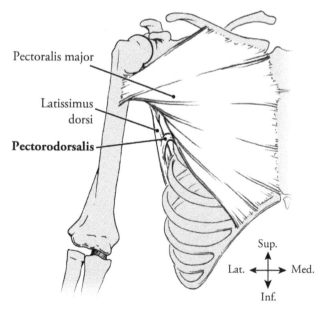

Figure 3.30. Anterior view of the pectoral region. Pectorodorsalis appears as a thin muscle belly from abdominal belly of pectoralis major to the latissimus dorsi. Anomalies labeled in bold (modified from Pettersen, 1979).

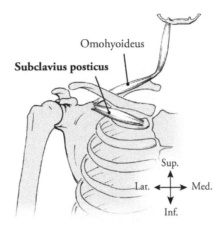

Figure 3.31. Anterior view of the pectoral region. A supernumerary subclavius posticus muscle originates from the first rib and inserts with the omohyoideus onto the scapula. Anomalies labeled in bold (modified from Ramirez-Castro and Bersu, 1978).

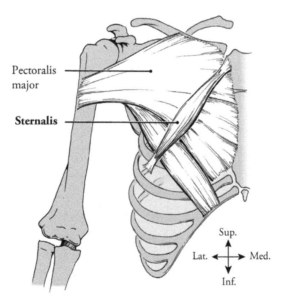

Figure 3.32. Anterior view of the pectoral region. A supernumerary sternalis muscle orginates from the manubrium and inserts onto lower ribs. Anomalies labeled in bold (modified from Ramirez-Castro and Bersu, 1978).

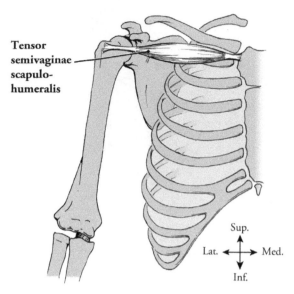

Tensor semivaginae scapulo-humeralis

Sup.

Lat. ←——→ Med.

Inf.

Figure 3.33. Anterior view of the pectoral region. Tensor semivaginae scapulohumeralis is present originating from the manubrium and inserting onto the humeral head. Anomalies labeled in bold (modified from Ramirez-Castro and Bersu, 1978).

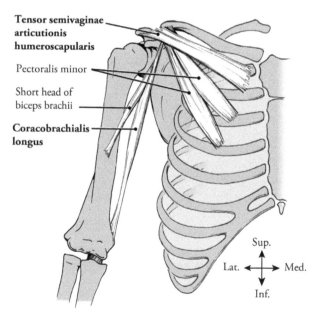

Tensor semivaginae articutionis humeroscapularis

Pectoralis minor

Short head of biceps brachii

Coracobrachialis longus

Sup.

Lat. ←——→ Med.

Inf.

Figure 3.34. Anterior view of the pectoral region. Both the coracobrachialis longus and tensor semivaginae articutionis humeroscapularis shown. Short head of the biceps brachii diminutive. Anomalies labeled in bold (modified from Aziz, 1979).

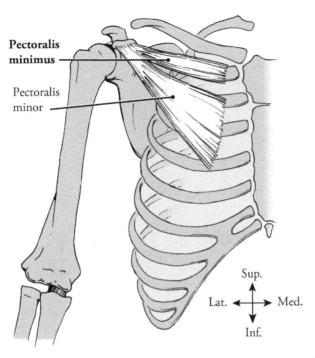

Figure 3.35. Anterior view of the pectoral region. Pectoralis minimus arises superior to the pectoralis minor and inserts onto the coracoid process. Anomalies labeled in bold (modified from Ramirez-Castro and Bersu, 1978).

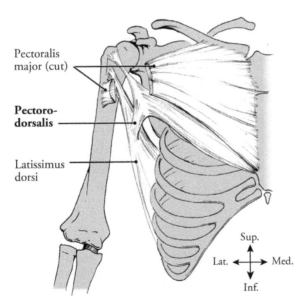

Figure 3.36. Anterior view of the pectoral region. Pectorodorsalis appears as a broad sheet from abdominal belly of pectoralis major to the latissimus dorsi. Anomalies labeled in bold (modified from Colacino and Pettersen, 1978).

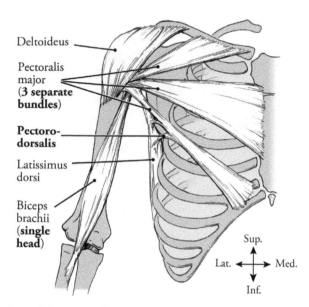

Figure 3.37. Anterior view of the pectoral region. Pectoralis major divided into three separate bundles. Pectorodorsalis originates from inferior belly. Biceps brachii has single head. Anomalies labeled in bold (Aziz, 1980).

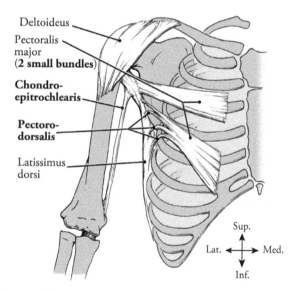

Figure 3.38. Anterior view of the pectoral region. Pectoralis major appears in two small bundles. Chondroepitrochlearis spans from latissimus dorsi to the medial epicondyle and pectorodorsalis appears in three bundles originating from the pectoralis major to the latissimus dorsi. Anomalies labeled in bold (modified from Aziz, 1979).

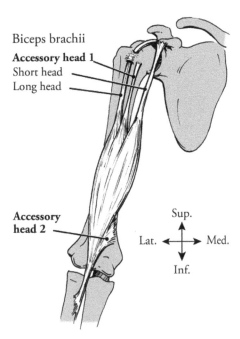

Figure 3.39. Anterior upper arm. Biceps brachii shown with two accessory heads. Anomalies labeled in bold (modified from Ramirez-Castro and Bersu, 1978).

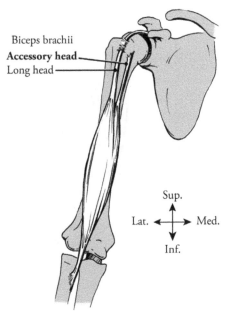

Figure 3.40. Anterior upper arm. Biceps brachii has one accessory head inserting onto the humeral head. Anomalies labeled in bold (modified from Aziz, 1981).

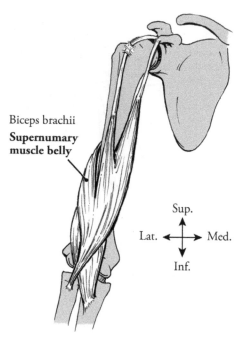

Biceps brachii
Supernumary muscle belly

Sup.

Lat. ◄──► Med.

Inf.

Figure 3.41. Anterior upper arm. Supernumerary head of the biceps brachii inserting around the humeral shaft. Anomalies labeled in bold (modified from Ramirez-Castro and Bersu, 1978).

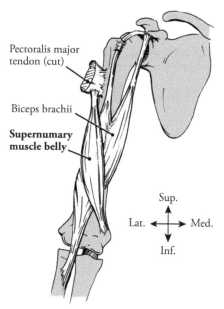

Pectoralis major tendon (cut)

Biceps brachii
Supernumary muscle belly

Sup.

Lat. ◄──► Med.

Inf.

Figure 3.42. Anterior upper arm. Supernumerary muscle crossing over normal biceps brachii to insert onto pectoralis major. Anomalies labeled in bold (modified from Ramirez-Castro and Bersu, 1978).

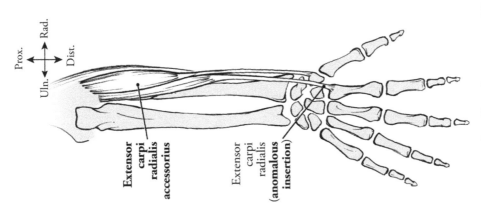

Figure 3.44. Forearm extensor view. Note the presence of the extensor carpi radialis accessorius and the anomalous insertion of the extensor carpi radialis. Anomalies labeled in bold (modified from Ramirez-Castro and Bersu, 1978).

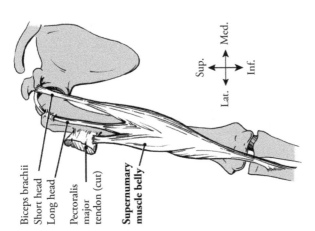

Figure 3.43. Anterior upper arm. Supernumerary muscle belly of the biceps brachii inserts onto pectoralis major. Anomalies labeled in bold (modified from Ramirez-Castro and Bersu, 1978).

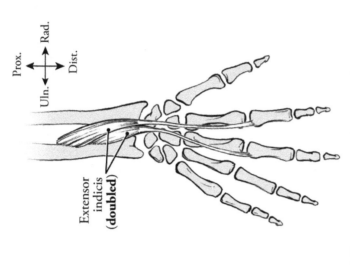

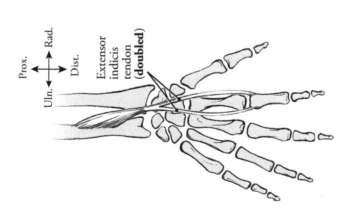

Figure 3.45. Forearm extensor view. Extensor indicis tendon doubled, but both tendons still insert onto digit one. Anomalies labeled in bold (modified from Ramirez-Castro and Bersu, 1978).

Figure 3.46. Forearm extensor view. Extensor indicis belly doubled. Muscles insert separately onto digits one and two. Anomalies labeled in bold (modified from Ramirez-Castro and Bersu, 1978).

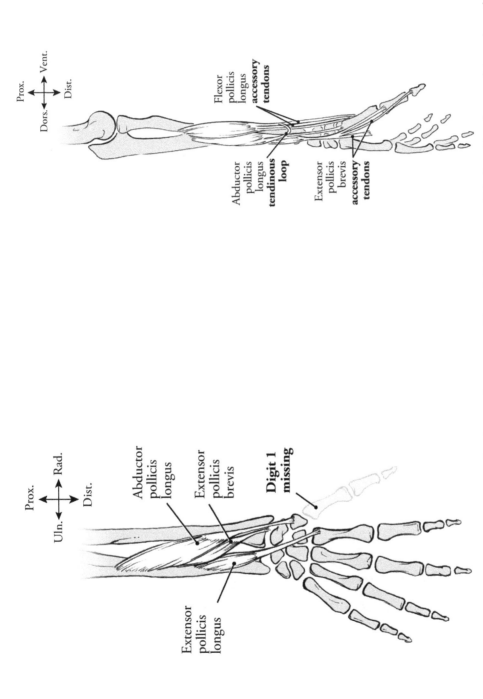

Figure 3.48. Forearm radial view. Thumb musculature gives of multiple anomalous tendons. Anomalies labeled in bold (modified from Ramirez-Castro and Bersu, 1978).

Figure 3.47. Forearm extensor view. Digit one is absent. Thumb musculature attaches to digit one and carpal bones. Anomalies labeled in bold (modified from Ramirez-Castro and Bersu, 1978).

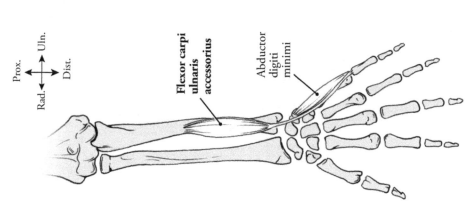

Figure 3.50. Forearm flexor view. A supernumerary flexor carpi ulnaris accessorius muscle arises from the ulna and inserts with the abductor digit minimi onto the proximal phalange of digit five. Anomalies labeled in bold (modified from Ramirez-Castro and Bersu, 1978).

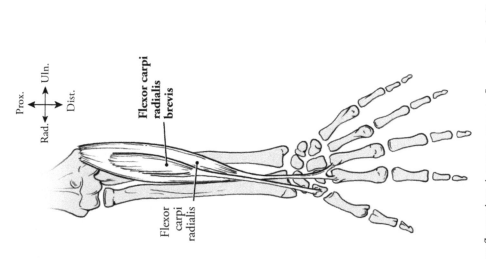

Figure 3.49. Forearm flexor view. A supernumerary flexor carpi radialis brevis muscle arises from the belly of the flexor carpi radialis and inserts onto the trapezium. Anomalies labeled in bold (modified from Ramirez-Castro and Bersu, 1978).

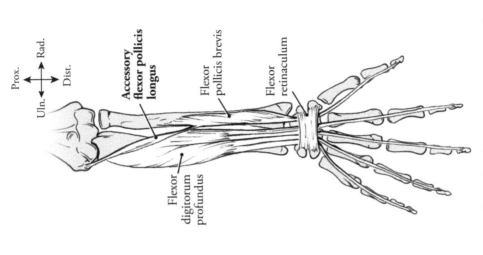

Figure 3.52. Forearm flexor view. A supernumerary accessory flexor pollicis longus muscle originates from the medial epicondyle and inserts onto the flexor pollicis longus. Anomalies labeled in bold (modified from Pettersen, 1979).

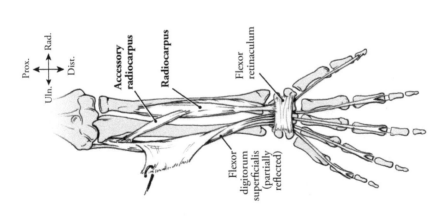

Figure 3.51. Forearm flexor view. A supernumerary radiocarpus muscle arises from the radius and inserts onto the flexor retinaculum. An accessory radiocarpus is also present originating from the flexor bellies and inserting onto the fibers of the radiocarpus. Anomalies labeled in bold (modified from Pettersen, 1979).

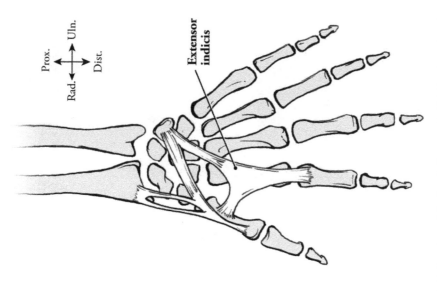

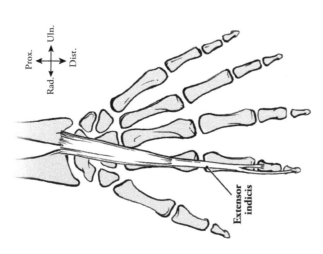

Figure 3.53. Posterior view of the hand. Note the distal origin of the extensor indicis and its distal doubled tendon. Anomalies labeled in bold (modified from Aziz, 1979).

Figure 3.54. Posterior view of the hand. Extensor indicis has broad tendon and also inserts onto the metacarpal of the first digit. Anomalies labeled in bold (modified from Colacino and Pettersen, 1978).

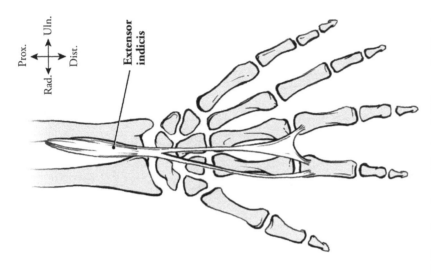

Figure 3.56. Posterior view of the hand. Extensor indicis gives off multiple tendons going to digits two and three. Anomalies labeled in bold (modified from Colacino and Pettersen, 1978).

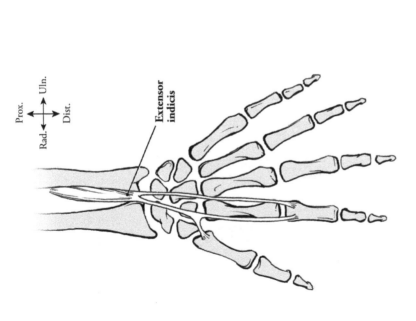

Figure 3.55. Posterior view of the hand. Extensor indicis gives off multiple tendons that insert onto both digits one and two. Anomalies labeled in bold (modified from Colacino and Pettersen, 1978).

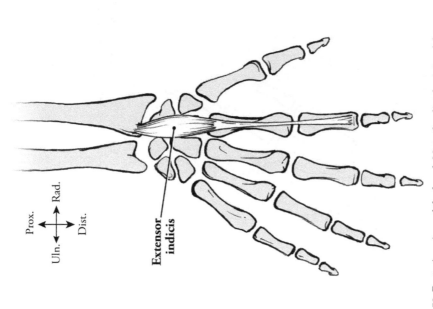

Figure 3.58. Posterior view of the hand. Note the distal origin of the extensor indicis and its diminutive size. Anomalies labeled in bold (modified from Colacino and Pettersen, 1978).

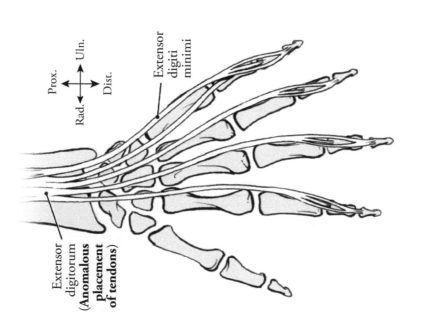

Figure 3.57. Posterior view of the hand. Anomalous placement of the tendons of the extensor digitorum and extensor digiti minimi to the ulnar sides of the metacarpophalangeal joints. Anomalies labeled in bold (modified from Ramirez-Castro and Bersu, 1978).

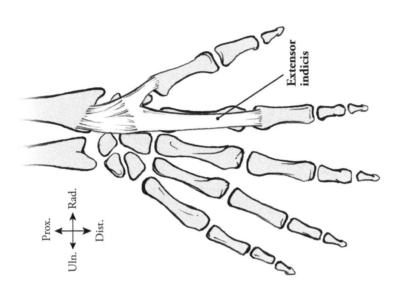

Figure 3.60. Posterior view of the hand. The extensor indicis has a broad tendon which inserts onto proximal phalange instead of the extensor expansion and gives off extra tendon to the metacarpal of digit one. Anomalies labeled in bold (modified from Colacino and Pettersen, 1978).

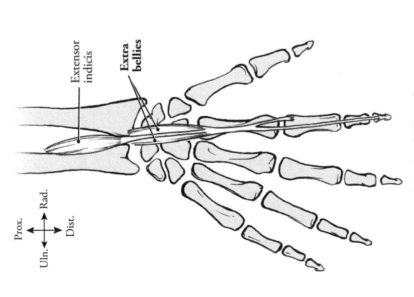

Figure 3.59. Posterior view of the hand. Note the extra supernumerary muscle bellies arising around the extensor indicis. Anomalies labeled in bold (modified from Colacino and Pettersen, 1978).

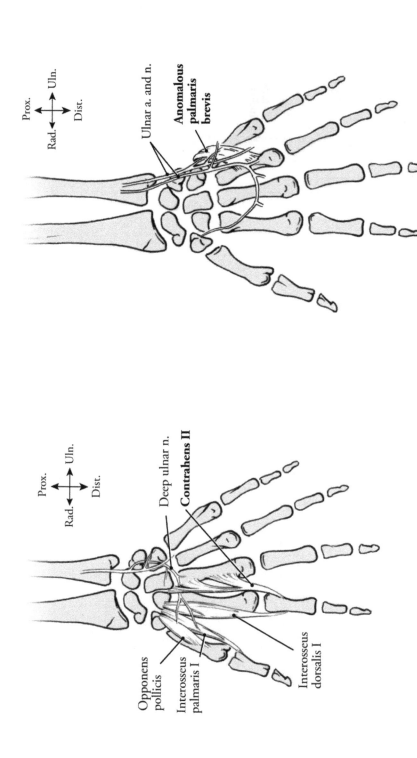

Figure 3.62. Palmar view of the hand. Note the anomalous palmaris brevis wrapping latero-proximal to medio-distal and laying deep to the ulnar artery and nerve. Anomalies labeled in bold (modified from Pettersen, 1979).

Figure 3.61. Palmar view of the hand. Note the presence of the contrahens II originating from the medial portion of metacarpal three and inserting onto ulnar side of digit two. Anomalies labeled in bold (modified from Dunlap, 1986).

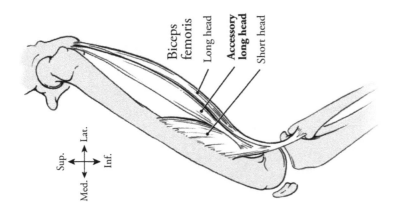

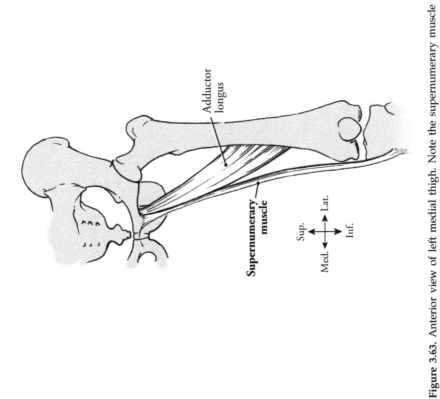

Figure 3.63. Anterior view of left medial thigh. Note the supernumerary muscle belly medial to the adductor longus. Anomalies labeled in bold (modified from Aziz, 1980).

Figure 3.64. Left lateral view of thigh. Biceps femoris has accessory long head just lateral to long head. Anomalies labeled in bold (modified from Pettersen, 1979).

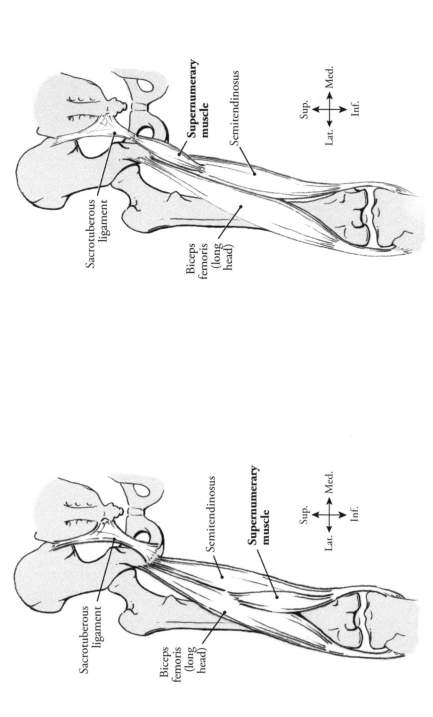

Figure 3.65. Posterior view of left thigh. A supernumerary muscle originates from the biceps femoris long head and inserts onto the semitendinosus tendon. Anomalies labeled in bold (modified from Ramirez-Castro and Bersu, 1978).

Figure 3.66. Posterior view of left thigh. A supernumerary muscle originates from the sacrotuberous ligament and inserts onto the semitendinosus belly. Anomalies labeled in bold (modified from Ramirez-Castro and Bersu, 1978).

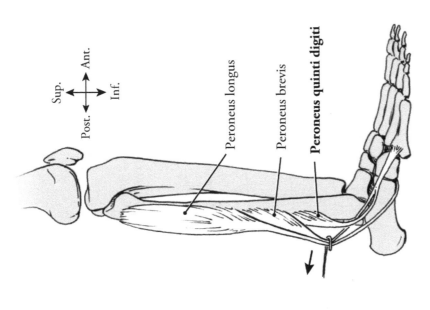

Figure 3.68. Lateral view of right leg. A supernumerary peroneus quinti digiti muscle originates from the distal fibula and inserts onto the peroneus brevis tendon. Anomalies labeled in bold (modified from Aziz, 1979).

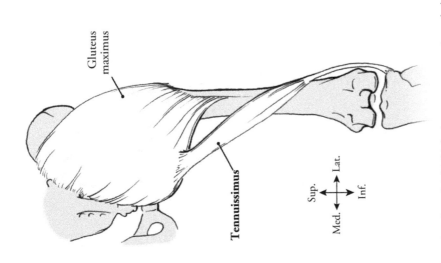

Figure 3.67. Posterior view of right thigh. A supernumerary tennuissimus muscle originates from the sacrotuberous ligament and fibers of the gluteus maximus to insert onto the lateral condyle of the tibia. Anomalies labeled in bold (modified from Ramirez-Castro and Bersu, 1978).

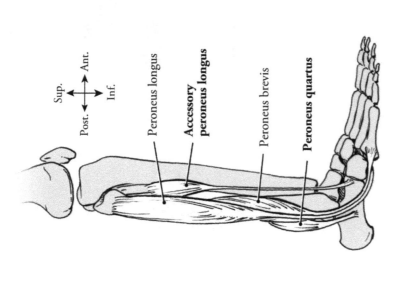

Figure 3.70. Lateral view of right leg. A supernumerary peroneus quartus muscle originates from the distal fibula and inserts onto the calcaneus. Note the presence of the accessory peroneus longus muscle. Anomalies labeled in bold (modified from Aziz, 1979).

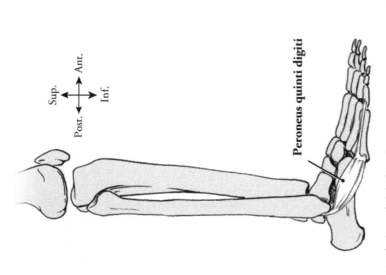

Figure 3.69. Lateral view of right leg. A supernumerary peroneus quinti digiti muscle originates from the distal fibula and inserts onto the metatarsal and proximal phalange of digit 5 of the foot. Anomalies labeled in bold (modified from Ramirez-Castro and Bersu, 1978).

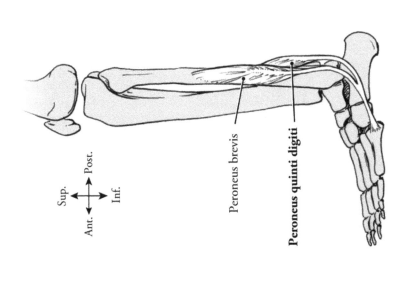

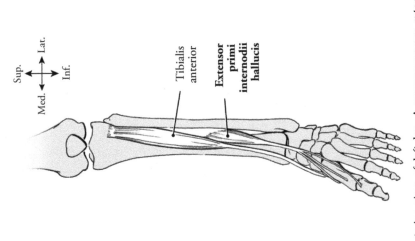

Figure 3.71. Anterior view of left leg. A supernumerary extensor primi internodii hallucis muscle originates from the proximal portion of tibia to metacarpal and proximal phalanx of hallux. Anomalies labeled in bold (modified from Aziz, 1979).

Figure 3.72. Lateral view of left leg. A supernumerary peroneus quinti digiti muscle originates from the distal fibula and inserts onto the peroneus brevis tendon. Anomalies labeled in bold (modified from Aziz, 1979)

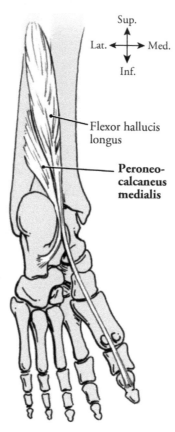

Figure 3.73. Plantar view of left foot. A supernumerary peroneocalcaneus medialis originates from the fibula and inserts onto the underside of the calcaneus. Anomalies labeled in bold (modified from Ramirez-Castro and Bersu, 1978).

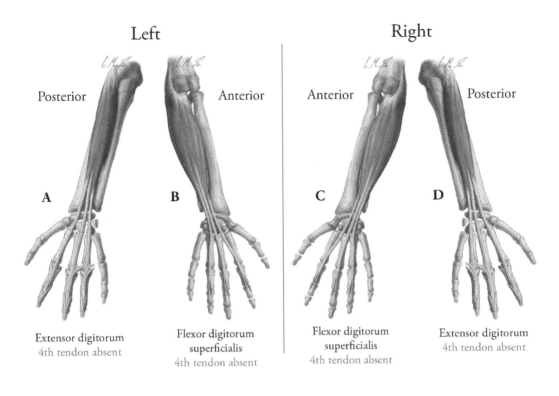

Left

Posterior

Anterior

Right

Anterior

Posterior

A

B

C

D

Extensor digitorum
4th tendon absent

Flexor digitorum
superficialis
4th tendon absent

Flexor digitorum
superficialis
4th tendon absent

Extensor digitorum
4th tendon absent

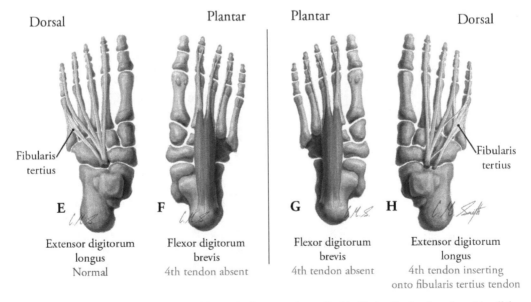

Dorsal

Plantar

Plantar

Dorsal

Fibularis
tertius

Fibularis
tertius

E

F

G

H

Extensor digitorum
longus
Normal

Flexor digitorum
brevis
4th tendon absent

Flexor digitorum
brevis
4th tendon absent

Extensor digitorum
longus
4th tendon inserting
onto fibularis tertius tendon

Figure 4.1. Trisomy 18 cyclopia upper and lower limb comparisons. Red indicates the tendon absent in all four limbs: 4th tendon to the 5th digit (**A-D, F, G**). Normal anatomy was observed on the dorsal surface of the left foot (**E**). The 4th tendon inserts onto separate fibularis tertius on the dorsal surface of the right foot (**H**).

Color image of this figure appears in the color plate section at the end of the book.

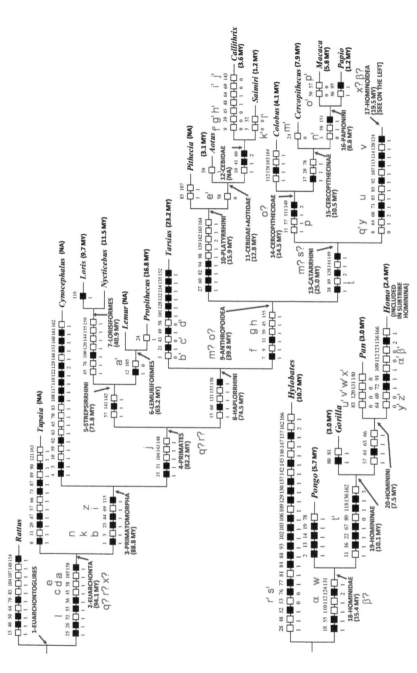

Figure 6.1. Single most parsimonious tree (L 301, CI 58, RI 73) obtained from the analysis of of 166 characters of the head, neck, pectoral and forelimb musculature (Diogo and Wood, 2011; 2012a). The unambiguous transitions that occurred in each branch are shown in white (homoplasic transitions) and black (non-homoplasic transitions) squares (numbers above and below the squares indicate the character and character state, respectively). Together with the name of each euarchontan clade is shown the respective estimate molecular divergence time, excepting for the genus *Homo* for which it is shown a time of origin exclusively based on the fossil record (see text for more details). A detailed description of the 28 unambiguous reversions to a plesiomorphic state is given in Diogo and Wood, 2012b; in the text of the present Chapter we only refer to some of these reversions (N.B., letters/symbols without a prime indicate the nodes where the respective original transitions from the plesiomorphic state to the derived state took place).

Color image of this figure appears in the color plate section at the end of the book.

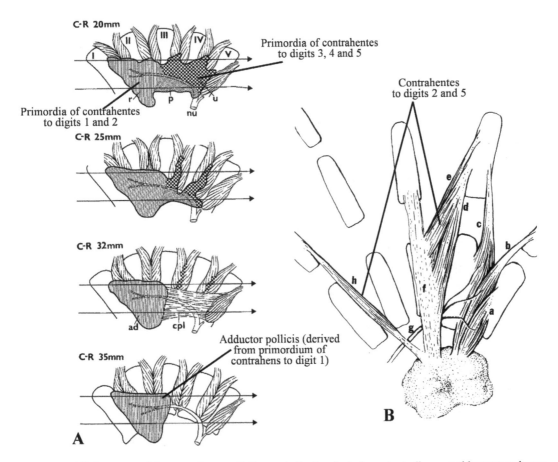

Figure 6.2. (A) Ontogeny of the contrahentes digitorum in the hand of a karyotypically normal human embryos showing how the contrahentes to digits other than digit 1 are usually lost (reabsorbed) early in development (modified from Cihak, 1972). Part of the interossei primordia (i.e., the flexores breves profundi layer) are shown between the metacarpals. r, u, p: radial, proximal and ulnar parts of contrahentes layer; nu: ulnar nerve; ad: adductor pollicis; cpl: contrahens plate; I-V: metacarpals I-V; CR: crown-rump length of the embryos. **(B)** Deep left hand musculature of a Trisomy 18 neonate (100 days old, female) showing the presence of contrahentes to digits 2 and 5 (the more superficial muscles, as well as the adductor pollicis, are now shown; modified from Dunlap et al., 1986). a: opponens pollicis; b: 'interosseous palmaris of Henle'; c: interosseous dorsalis 1; d: contrahens to digit 2; e: interosseous palmaris 1; f: contrahens fascia/medial raphe; g: deep branch of ulnar nerve; h: contrahens to digit 5.

References

Akita, K., T. Shimokawa and T. Sato. 2000. Aberrant muscle between the temporalis and the lateral pterygoid muscles: M. pterygoideus proprius (Henle). *Clin. Anat.*, 14: 288–291.

Alberch, P. 1989. The logic of monsters: evidence for internal constraint in development and evolution. *Geobios, Mém. Spécial*, 12: 21–57.

Aldridge, K., R.H. Reeves, L.E. Olson and J.T. Richtsmeier. 2007. Differential effects of trisomy on brain shape and volume in related aneuploid mouse model. *Am. J. Med. Genet.*, A 143: 1060–1070.

Allanson, J.E., P. O'Hara, L.G. Farkas and R.C. Nair. 1993. Anthropometric craniofacial pattern profiles in Down syndrome. *Am. J. Med. Genet.*, 47: 748–752.

Ammendolia, A. 2008. Extensor digitorum brevis manus associated with a dorsal wrist ganglion: Case report. *Clin. Anat.*, 21: 794–795.

Anatomy Atlases. 2014. http://www.anatomyatlases.org/atlasofanatomy/index.shtml.

Arnold, C., L.J. Matthews and C.L. Nunn. 2010. The 10kTrees Website: a new online resource for primate phylogeny. *Evol. Anthropol.*, 19: 114–118.

Arráez-Aybar, L.A., J. Sobrado-Perez and J.R. Merida-Velasco. 2003. Left musculus sternalis. *Clin. Anat.*, 16: 350–354.

Asha, K.R., S. Lakshmiprabha, C.M. Nanjaish and S.N. Prashanth. 2011. Craniofacial anthropometric analysis in Down syndrome. *Indian J. Pediatrics*, 78: 1091–1095.

Attanasio, C., A.S. Nord, Y. Zhu, M.J. Blow, Z. Li, D.K. Liberton, H. Morrison, I. Plajzer-Frick, A. Holt, R. Hosseini, S. Phouanenavong, J.A. Akiyama, M. Shoukry, V. Afzal, E.M. Rubin, D.R. FitzPatrick, B. Ren, B. Hallgrimsson, L.A. Pennacchio and A. Visel. 2013. Fine tuning of craniofacial morphology by distant-acting enhancers. *Science*, 342: 6157.

Aziz, M.A. 1979. Muscular and other abnormalities in a case of Edwards' syndrome (18-Trisomy). *Teratology*, 20: 303–312.

Aziz, M.A. 1980. Anatomical defects in a case of Trisomy 13 with a D/D translocation. *Teratology*, 22: 217–227.

Aziz, M.A. 1981a. Muscular anomalies caused by delayed development in human aneuploidy. *Clin. Gen.*, 19: 111–116.

Aziz, M.A. 1981b. Possible atavistic structures in human aneuploids. *Am. J. Phys. Anthropol.*, 54: 347–353.

Aziz, M.A. and S.S. Dunlap. 1986. The human extensor digitorum profundus muscle with comments on the evolution of the primate hand. *Primates*, 27: 293–319.

Aziz, M.A. and J.C. McKenzie. 1999. The dead can still teach the living: The status of cadaver-based Anatomy in the age of electronic media. *Perspect. Biol. Med.*, 402: 421.

Bagic, I. and Z. Verzak. 2003. Craniofacial anthropometric analysis in Down's syndrome patients. *Collegium Antropolog.*, 27: 23–30.

Barash, B.A., L. Freedman and J.M. Optizj. 1970. Anatomic studies in the 18-Trisomy syndrome. *Birth Def.*, 4: 3–15.

Baxter, L.L., T.H. Moran, J.T. Richtsmeier, J. Troncoso and R.H. Reeves. 2000. Discovery and genetic localization of Down syndrome cerebellar phenotypes using the Ts65Dn mouse. *Hum. Mol. Genet.*, 9: 195–202.

Belichenko, N.P., P.V. Belichenko, A.M. Kleschevnikov, A. Salehi, R.H. Reeves and W.C. Mobley. 2009. The "Down syndrome critical region" is sufficient in the mouse model to confer behavioral, neuro-physiological, and synaptic phenotypes characteristic of Down syndrome. *J. Neurosci.*, 29: 5938–5948.

Bergman, R., S.A. Thompson, A. Afifi and F. Saadeh. 1988. *Compendium of Human Anatomic Variations: Text, Atlas and World Literature*. Baltimore: Urban & Schwarzenburg.

Bersu, E.T. and J.L. Ramirez-Castro. 1977. Anatomical analysis of the developmental effects of aneuploidy in Man—the 18-Trisomy Syndrome: I. Anomalies of the head and neck. *Am. J. Med. Gen.*, 1: 173–193.

Bersu, E.T. 1980. Anatomical analysis of the developmental effects of aneuploidy in Man—The Down syndrome. *Am. J. Med. Gen.*, 5: 399–420.

Billingsley, C.N., J.A. Allen, S.L. Deitz, J.D. Blazek, D.D. Baumann, A. Newbauer, A. Darrah, B.C. Long, B. Young, M. Clement, R.W. Doerge and R.J. Roper (In press). Non-trisomic Homeobox gene expression during craniofacial development in the Ts65Dn mouse model of Down syndrome. *Am. J. Med. Genet.*

Blazek, J.D., C.N. Billingsley, A. Newbauer and R.J. Roper. 2010. Embryonic and not maternal trisomy causes developmental attenuation in the Ts65Dn mouse model for Down syndrome. *Dev. Dyn.*, 239: 1645–1653.

Blazek, J.D., A. Gaddy, R. Meyer, R.J. Roper and J. Li. 2011. Disruption of bone development and homeostasis by trisomy in Ts65Dn Down syndrome mice. *Bone*, 48: 275–280.

Bosma, J.F. 1986. *Anatomy of the Infant Head*. Johns Hopkins University Press.

Blaas, H.K., A.G. Eriksson, K.Å. Salvesen, C.V. Isaksen and B. Christensen. 2002. Brains and faces in holoprosencephaly: Pre- and postnatal description of 30 Cases. *Ultrasound Obstet. Gynecol.*, 19: 24–38.

Blitz, E., A. Sharir, H. Akiyama and E. Zelzer. 2013. Tendon-bone attachment unit is formed modularly by a distinct pool of *Scx*- and *Sox9*-positive progenitors. *Development*, 140: 2680–2690.

Bober, E., T. Franz, H. Arnold, P. Gruss and P. Tremblay. 1994. *Pax-3* is required for the development of limb muscles: A possible role for the migration of dermomyotomal muscle. *Progenitor Cells*, 612: 603–612.

Bonastre, V., M. Rodríguez-Niedenführ, D. Choi and J.R. Sañudo. 2002. Coexistence of a pectoralis quartus muscle and an unusual axillary arch: Case report and review. *Clin. Anat.*, 15: 366–370.

Bookstein, F.L., P. Gunz, P. Mitterœcker, H. Prossinger, K. Schaefer and H. Seidler. 2003. Cranial integration in *Homo*: Singular warps analysis of the midsagittal plane in ontogeny and evolution. *J. Human Evolution*, 44: 167–187.

Bowler, P.J. 2007. *Fins and limbs and fins into limbs: The historical context, 1840–1940*. In *Fins into Limbs. Evolution, Development, and Transformation* (Hall B.K. Ed.), 7–14. The University of Chicago Press, Chicago and London.

Bowler, P. J. 1996. *Life's splendid Drama: Evolutionary Biology and the Reconstruction of Life's Ancestry, 1860–1940*. The University of Chicago Press, Chicago.

Brohmann, H., K. Jagla and C. Birchmeier. 2000. The role of *Lbx1* in migration of muscle precursor cells. *Development*, 127: 437–445.

Bufill, E., J. Agusti and R. Blesa. 2011. Human neoteny revisited: the case of synaptic plasticity. *Am. J. Hum. Biol.*, 23: 729–739.

Buckingham, M., L. Bajard, T. Chang, P. Daubas, J. Hadchouel, S. Meilhac, D. Montarras, D. Rocancourt and F. Relaix. 2003. The Formation of Skeletal Muscle: From Somite to Limb. *J. Anat.*, 202: 59–68.

Bugge, M., A. Collins, M.B. Petersen, J. Fisher, C. Brandt, J.M. Hertz, L. Tranebjaerg, C. de Lozier-Blanchet, P. Nicolaides, K. Brøndum-Nielsen, N. Morton and M. Mikkelsen. 1998. Non-Disjunction of Chromosome 18. *Human Mol. Gen.*, 7: 661–669.

Burrows, A.M., B.M. Waller, L.A. Parr and C.J. Bonar. 2006. Muscles of facial expression in the Chimpanzee (*Pan troglodytes*): Descriptive, comparative and phylogenic contexts. *J. Anat.*, 208: 153–167.

Burrows, A.M., B.M. Waller and L.A. Parr. 2009. Facial musculature in the Rhesus Macaque (*Macaca mulatta*): Evolutionary and functional contexts with comparison to Chimpanzees and Strepsirrhines. *J. Anat.*, 215: 320–34.

Caputo, V., B. Lanza and R. Palmieri. 1995. Body elongation and limb reduction in the genus *Chalcides* Laurenti 1768 (Squamata Scincidae): a comparative study. *Trop. Zool.*, 8: 95–152.

Carvalho, R.L. and D.A. Vasconcelos. 2011. Motor behavior in Down syndrome: atypical sensoriomotor control. In *Prenatal diagnosis and screening for Down Syndrome* (Deu, S. Ed.), 33–42. Rejeka: In Tech.

Castilla, E.E., R.L. da Fonseca, M. da Graça Dutra, E. Bermejo, L. Cuevas and M.L. Martínez-Frías. 1996. Epidemiological analysis of rare polydactylies. *Am. J. Med. Genet.*, 65: 295–303.

Cereda, A. and J.C. Carey. 2012. The Trisomy 18 Syndrome. *Orphanet Journal of Rare Diseases*, 7: 81.

Cihak, R. 1972. Ontogenesis of the skeleton and intrinsic muscles of the human hand and foot. *Adv. Anat. Embryol. Cell. Biol.*, 46: 1–194.

Chakrabarti, L., Z. Galdzicki and T.F. Haydar. 2007. Defects in embryonic neurogenesis and initial synapse formation in the forebrain of the Ts65Dn mouse model of Down syndrome. *J. Neurosci.*, 27: 11483–11495.

Chen, Y., N.C. Ozturk, L. Ni, C. Goodlett and F.C. Zhou. 2011. Strain differences in developmental vulnerability to alcohol exposure via embryo culture in mice. *Alcohol Clin. Exp. Res.*, 35: 1293–1304.

Chudzinski, T. 1896. *Quelques observations sur les muscls peauciers du crâne et de la face dans les races humaines*. Paris.

Chung, K. 2012. *Hand and Upper Extremity Reconstruction with DVD: A Volume in the Procedures in Reconstructive Surgery Series*. Philadelphia: Saunders Publ.

Chung, I.-H., J. Han, J. Iwata and Y. Chai. 2010. *Msx1* and *Dlx5* function synergistically to regulate frontal bone development. *Genesis*, 48: 645–655.

Cleves, M.A., C.A. Hobbs, P.A. Cleves, J.M. Tilford, T.M. Bird and J.M. Robbins. 2007. Congenital defects among liveborn infants with Down syndrome. *Birth Defects Res.*, 79: 657–663.

Coates,M.I., and J.A. Clack. 1990. Polydactyly in the earliest known. *Nature*, 347: 66–69.

Coates, M.I., J.E. Jeffery and M. Ruta. 2002. Fins to limbs: what the fossils say. *Evol. Dev.*, 4: 390–401.

Colacino, S. and J. Pettersen. 1978. Analysis of the gross anatomical variations found in four cases of Trisomy 13. *Am. J. Med. Gen.*, 2: 31–50.

Costa, A.C., K. Walsh and M.T. Davisson. 1999. Motor dysfunction in a mouse model for Down syndrome. *Physiol. Behav.*, 68: 211–220.

Cowley, P.M., S. Keslacy, F.A. Middleton, L.R. DeRuisseau, B. Fernhall, J.A. Kanaley and K.C. DeRuisseau. 2012. Functional and biochemical characterization of soleus muscle in Down syndrome mice: insight into the muscle dysfunction seen in the human condition. *Am. J. Physiol. Regul. Integr. Comp. Physiol.*, 303: 1251–1260.

Darwin, C. 1859. *On the origin of species.* London. John Murray.

Darwin, C. 1871. *The descent of man, and selection in relation to sex.* London. John Murray.

Darwin, C. 1872. *The expression of the emotions in man and animals.* London. John Murray.

Demyer, W., W. Zeman and C.G. Palmer. 1964. The face predicts the brain: Diagnostic significance of median facial anomalies for holoprosencephaly (arhinencephaly). *Pediatrics*, 34: 256–63.

Deitz, S.L. and R.J. Roper. 2011. Trisomic and allelic differences influence phenotypic variability during development of Down syndrome mice. *Genetics*, 189: 1487–1495.

Diogo, R. 2007. *On the origin and evolution of higher-clades: osteology, myology, phylogeny and macroevolution of bony fishes and the rise of tetrapods.* Enfield: Science Publishers.

Diogo, R. 2010. Comparative anatomy, anthropology and archaeology as case studies on the influence of human biases in natural sciences: The origin of 'Humans', of 'Behaviorally modern humans' and of 'Fully civilized humans'. *Open Anat. J.*, 2: 86–97.

Diogo, R. and V. Abdala. 2010. *Muscles of Vertebrates—Comparative Anatomy, Evolution, Homologies and Development.* Taylor and Francis, Oxford.

Diogo, R., V. Abdala, N.L. Lonergan and B. Wood. 2008a. From fish to modern humans—comparative anatomy, homologies and evolution of the head and neck musculature. *J. Anat.*, 213: 391–424.

Diogo, R., V. Abdala, M.A. Aziz, N.L. Lonergan and B. Wood. 2009a. From fish to modern humans—comparative anatomy, homologies and evolution of the pectoral and forelimb musculature. *J. Anat.*, 214: 694–716.

Diogo, R., Y. Hinits and S. Hughes. 2008b. Development of mandibular, hyoid and hypobranchial muscles in the zebrafish, with comments on the homologies and evolution of these muscles within bony fish and tetrapods. *BMC Dev. Biol.* 8: 24–46.

Diogo, R., M. Linde-Medina, V. Abdala and M. Ashley-Ross. 2013a. New, puzzling insights from comparative myological studies on the old and unsolved forelimb/hindlimb enigma. *Biolog. Rev.*, 88: 196–214.

Diogo, R., P. Murawala and E.M. Tanaka. 2013b. Is salamander hindlimb regeneration similar to that of the forelimb? Anatomical and morphogenetic analysis of hindlimb muscle regeneration in GFP-transgenic axolotls as a basis for regenerative and developmental studies. *J. Anat.*, 10: 459–68.

Diogo, R. and J.L. Molnar. 2014. Comparative anatomy, evolution and homologies of the tetrapod hindlimb muscles, comparisons with forelimb muscles, and deconstruction of the forelimb-hindlimb serial homology hypothesis. *Anat. Rec.*, 297: 1047–1075.

Diogo, R., F. Pastor, F. De Paz, J.M. Potau, G. Bello-Hellegouarch, E.M. Ferrero and R. Fischer. 2012b. The head and neck muscles of the serval and tiger: homologies, evolution and proposal of a mammalian and a veterinary muscle ontology. *Anat. Rec.*, 295: 2157–2178.

Diogo, R., J.M. Potau, J.F. Pastor, F.J. de Paz, E.M. Ferrero, G. Bello, M. Barbosa and B. Wood. 2010. *Photographic and Descriptive Musculoskeletal Atlas of Gorilla.* Taylor & Francis, Oxford.

Diogo, R., B.G. Richmond and B.A. Wood. 2012a. Evolution and homologies of modern human hand and forearm muscles, with notes on thumb movements and tool use. *J. Hum. Evol.*, 63: 64–78.

Diogo, R. and E.M. Tanaka. 2014. Development of fore- and hindlimb muscles in GFP-transgenic axolotls: Morphogenesis, the tetrapod bauplan, and new insights on the forelimb-hindlimb Enigma. *J. Exp. Zool.*, 322B: 106–127.

Diogo, R., S. Walsh, C.M. Smith, J.M. Ziermann, V. Abdala and J.R. Hutchinson (in press). Towards the resolution of a long-standing evolutionary question: identity and attachments are mainly related to topological position and not to anlage or homeotic identity of digits. *J Anat.*

Diogo, R. and B.A. Wood. 2010. Human muscular variations: comparative, evolutionary and developmental perspectives. *FASEB J.*, 24: 61.

Diogo, R. and B.A. Wood. 2011. Soft-tissue anatomy of the primates: Phylogenetic analyses based on the muscles of the head, neck, pectoral region and upper limb, with notes on the evolution of these muscles. *J. Anat.*, 219: 273–359.

Diogo, R. and B.A. Wood. 2012a. *Comparative Anatomy and Phylogeny of Primate Muscles and Human Evolution.* CRC press.

Diogo, R. and B.A. Wood. 2012b. Violation of Dollo's law: evidence of muscle reversions in primate phylogeny and their implications for the understanding of the ontogeny, evolution and anatomical variations of modern humans. *Evolution*, 66: 3267–3276.

Diogo, R. and B.A. Wood. 2013. The broader evolutionary lessons to be learned from a comparative and phylogenetic analysis of primate muscle morphology. *Biol. Rev.*, 88: 988–1001.

Diogo, R., B.A. Wood, M.A. Aziz and A. Burrows. 2009b. On the origin, homologies and evolution of primate facial muscles, with a particular focus on hominoids and a suggested unifying nomenclature for the facial muscles of the Mammalia. *J. Anat.*, 215: 300–319.

Diogo, R. and J.M. Ziermann. 2014. Development of fore- and hindlimb muscles in frogs: morphogenesis, homeotic transformations, digit reduction, and the forelimb-hindlimb enigma. *J. Exp. Zool. B. (Mol. Dev. Evol.)*, 322: 86–105.

Drake, R.L., A.W. Vogl and A.W.M. Mitchel. 2014. *Gray's Anatomy for Students 2nd Edt.* Philadelphia: Churchill, Livings Tone, Elsevier.

Dunlap, D.G. 1967. The development of the musculature of the hindlimb in the frog, *Rana pipiens. J. Morphol.*, 119: 241–258.

Dunlap, S.S., M.A. Aziz and K.N. Rosenbaum. 1986. Comparative anatomical analysis of human Trisomies 13, 18, and 21: I. The forelimb. *Teratology*, 33: 159–86.

Dunlevy, L., M. Bennett, A. Slender, E. Lana-Elola, V.L. Tybulewicz, E.M. Fisher and T. Mohun. 2010. Down's syndrome-like cardiac developmental defects in embryos of the transchromosomic Tc1 mouse. *Cardiovasc. Res.*, 88: 287–295.

Duprez, D. 2002. Signals regulating muscle formation in the limb during embryonic development. *Int. J. Dev. Biol.*, 46: 915–925.

El-Naggar, M.M. and F.I. Zahir. 2001. Two bellies of the coracobrachialis muscle associated with a third head of biceps brachii muscle. *Clin. Anat.*, 14: 379–382.

El-Naggar, M.M. and S. Al-Saggaf. 2004. Variant of the coracobrachialis muscle with a tunnel for the median nerve and brachial artery. *Clin. Anat.*, 17: 139–143.

Elsayed, S.M. and G.M. Elsayed. 2009. Phenotype of apoptopic lymphocytes in children with Down syndrome. *Immun. Ageing*, 6: 2.

Epstein, C.J. 2001. Down syndrome (Trisomy 21). In *The Metabolic and Molecular Bases of Inherited Disease* (Scriver, C.R., A.L. Beaudet, W.S. Sly and D. Valle Eds.). McGraw-Hill, New York. pp. 1223–1256.

Epstein, D. 2013. *The sports gene inside the science of extraordinary athletic performance.* New York: The Penguin Press.

Escorihuela, R.M., A. Fernandez-Teruel, I.F. Vallina, C. Baamonde, A. Lumbreras, M. Dierssen, A. Tobena and J. Florez. 1995. A behavioral assessment of Ts65Dn mice: a putative Down syndrome model. *Neurosci. Lett.*, 199: 143–146.

Fabre, P.H., A. Rodrigues and E.J.P. Douzery. 2009. Patterns of macroevolution among primates inferred from a supermatrix of mitochondrial and nuclear DNA. *Mol. Phylogen. Evol.*, 53: 808–25.

Fabrezi, M., V. Abdala and M.I.M. Oliver. 2007. Developmental basis of limb homology in lizards. *Anat Rec.*, 290: 900–912.

Fabrizio, P.A. and F.R. Clemente. 1997. Variation in the triceps brachii muscle: A fourth muscular head. *Clin. Anat.*, 10: 259–263.

Forcada, P., M. Rodríguez-Niedenführ, M. Llusá and A. Carrera. 2001. Subclavius posticus muscle: Supernumerary muscle as a potential cause for Thoracic Outlet Syndrome. *Clin. Anat.*, 14: 55–57.

Francis-west, P., R. Ladher, A. Barlow and A. Graveson. 1998. Signalling Interactions during Facial Development. *Mechanisms of Development*, 75: 3–28.

Fujinaga, M., T.H. Shepard and J. Fitzsimmons. 1990. Trisomy 13 in the fetus. *Teratology*, 41: 233–238.

Futamura, R. 1906. Uber die Entwicklung der Facialismuskulatur des Menschen Facialismuskulatur des Menschen. *Anat. Hefte. Bd*, 30: 433–460.

Futamura, R. 1907. Beitrage zur vergleichenden Entwicklungsgeschichte der Facialismukulatur. *Anat. Hefe. Bd*, 32: 479–500.

Galis, F. and J.A.J. Metz. 2007. Evolutionary novelties: the making and breaking of pleiotropic constraints. *Int. Comp. Biol.*, 47: 409–419.

Garzozi, H.J. and S. Barkay. 1985. Case of True Cyclopia. *British J. Ophthal.*, 69: 307–311.

Gasser, R. 1967. The development of facial muscles in Man. *Am. J. Anat.*, 120: 357–376.

Georgiev, G.P., L. Jelev and L. Surchev. 2007. Axillary arch in Bulgarian population: Clinical significance of the arches. *Clin. Anat.*, 20: 286–291.

Gilbert, S. 2000. *Developmental Biology: 6th Edition.* Massachusetts: Sinauer Associates.

Gilsanz, V. and O. Ratib. 2005. *Hand Bone Age: A Digital Atlas of Skeletal Maturity.* Springer.

Gotlieb, N. R. 2009. *Quantification and analysis of apoptosis in embryonic atrioventricular endocardial cushions of the Ts65Dn mouse model for Down Syndrome.* Graduate Thesis, Franklin and Marshall College.

Gould, S.J. 1977. *Ontogeny and Phylogeny.* Harvard University Press, Cambridge.

Gould, S.J. 1993. *The Mismeasure of Man.* New York, W W Norton & Co Inc.

Gould, S.J. 2002. *The Structure of Evolutionary Theory.* Belknap, Harvard.

Grenier, J., M.A. Teillet, R. Grifone, R.G. Kelly and D. Duprez. 2009. Relationship between neural crest cells and cranial mesoderm during head muscle development. *PLOS ONE 4.* doi: 10.1371/journal.pone.0004381.

Gross, M.K., L. Moran-Rivard, T. Velasquez, M.N. Nakatsu, K. Jagla and M. Goulding. 2000. *Lbx1* is required for muscle precursor migration along a lateral pathway into the limb. *Development,* 127: 413–24.

Guidi, S., P. Bonasoni, C. Ceccarelli, D. Santini, F. Gualtieri, E. Ciani and R. Bartesaghi. 2008. Neurogenesis impairment and increased cell death reduce total neuron number in the hippocampal region of fetuses with Down syndrome. *Brain Pathol.,* 18: 180–197.

Hall, B.K. and J.A. Gillis. 2013. Incremental evolution of the neural crest, neural crest cells and neural crest-derived skeletal tissues. *J. Anat.,* 222: 19–31.

Hallgrímsson, B., K. Willmore and B.K. Hall. 2002. Canalization, developmental stability, and morphological integration in primate limbs. *American Journal of Physical Anthropology Suppl.,* 35: 131–158.

Harel, I., Y. Maezawa, R. Avraham, A. Rinon, H. Ma, J.W. Cross and N. Leviatan. 2011. Pharyngeal mesoderm regulatory network controls cardiac and head muscle morphogenesis. *Cardiovascular Research 91,* 46: 196–202.

Harry, W.G., J.D. Bennett and S.C. Guha. 1997. Scalene muscles and the brachial plexus: Anatomical variations and their clinical significance. *Clin. Anat.,* 10: 250–252.

Hawli, Y., M. Nasrallah and G. El-Hajj Fuleihan. 2009. Endocrine and musculoskeletal abnormalities in patients with Down syndrome. *Nat. Rev. Endocrinol.,* 5: 327–334.

Heiss, H. 1957. Beiderseitige kongenitale daumenlose Fünffingerhand bei Mutter und Kind. *Z. Anat. Entw-Gesch.,* 120: 226–231.

Hill, E. 1920. Cyclopia, it's bearing upon certain problems of teratogensis and of normal embryology; with a description of a cyclocephalic monster. *Trans. Am. Ophthalm. Soc.,* 18: 329–384.

Huber, E. 1931. *Evolution of facial musculature and facial expression. Baltimore:* The Johns Hopkins Press.

Innis, J.W., E.H. Margulies and S. Kardia. 2002. Integrative biology and the developing limb bud 1. *Evol. Dev.,* 389: 378–389.

Jay, F.A., N.A. Jacobson and Q.A. Fogg. 2008. Anatomical variations of the plantaris muscle and a potential role in Patellofemoral Pain Syndrome. *Clin. Anat.,* 21: 178–181.

Jones, L.C. 1979. The morphogenesis of the thigh of the mouse with special reference to tetrapod muscle homologies. *J. Morphol.,* 162: 275–310.

Kardon, G. 1998. Muscle and tendon morphogenesis in the avian hind limb. *Development,* 125: 4019–4032.

Keyte, A.L. and K.K. Smith. 2010. Developmental origins of precocial forelimbs in marsupial neonates. *Development,* 137: 4283–4294.

Kida, M.Y., A. Izumi and S. Tanaka. 2000. Sternalis muscle: Topic for debate. *Clin. Anat.,* 13: 138–140.

Kish, P.E., B.L. Bohnsack, D. Gallina, D.S. Kasprick and A. Kahana. 2011. The eye as an organizer of craniofacial development. *Genesis,* 49: 222–30.

Kjaer, I., J.W. Keeling and N. Graem. 1991. The midline craniofacial skeleton in holoprosencephalic fetuses. *J. Med. Gen.,* 28: 846–855.

Kjaer, I., J.W. Keeling, N.M. Smith and B.F. Hansen. 1997. Pattern of malformations in the axial skeleton in human triploid fetuses. *Am. J. Med. Gen.,* 72: 216–221.

Kobayashi, N., S. Saito, H. Wakisaka and S. Matsuda. 2003. Anomalous flexor of the little finger. *Clin. Anat.,* 16: 40–43.

Kohn, L.T., J.M. Corrigan and M.S. Donaldson. 2000. *To Err Is Human: Building a Safer Health System.* Washington, DC. National Academies Press.

Köntges, G. and A. Lumsden. 1996. Rhombencephalic neural crest segmentation is preserved throughout craniofacial ontogeny. *Development,* 122: 3229–3242.

Korenberg, J.R., X.N. Chen, R. Schipper, Z. Sun, R. Gonsky, S. Gerwehr, N. Carpenter, C. Daumer, P. Dignan and C. Disteche. 1994. Down syndrome phenotypes: the consequences of chromosomal imbalance. *PNAS,* 91: 4997–5001.

Lang, A.P., M. Schlager and H.A. Gardner. 1976. Trisomy 18 and cyclopia. *Teratology,* 14: 195–204.

Laurin, M. 2011. Paleontological evidence: Origin and early evolution of limbed vertebrates. In *How Vertebrates moved onto Land* (Bels, V., A. Casinos, J. Davenport, J.P. Gasc, M. Jamon, M. Laurin and S. Renous Eds.), 33–73. Museum national d'Histoire naturelle, Paris.

Le Double, A.F. 1897. *Traité des variations du système musculaire de l'homme.* Librairie C. Reinwald, Paris.

Leroi, A.M. 2005. *Mutants: On Genetic Variety and the Human Body.* Penguin Books.

Li, Z., T. Yu, M. Morishima, A. Pao, J. Laduca, J. Conroy, N. Nowak, S. Matsiu, I. Shiraishi and Y.E. Yu. 2007. Duplication of the entire 22.9 Mb human chromosome 21 syntenic region on mouse Chromosome 16 causes cardiovascular and gastrointestinal abnormalities. *Hum. Mol. Genet.*, 16: 1359–1366.

Light, T.R. 1992. Thumb reconstruction. *Hand Clin.*, 8: 161–175.

Lightoller, G.S. 1928. The facial muscles of three orang utans and two Cercopithecidae. *J. Ant.*, 63: 19–35.

Liu, C., M. Morishima, T. Yu, S. Matsui, L. Zhang, D. Fu, A. Pao, A.C. Costa, K.J. Gardiner, J.K. Cowell, N.J. Nowak, M.S. Parmacek, P. Liang, A. Baldini and Y.E. Yu. 2011. Genetic analysis of Down syndrome-associated heart defects in mice. *Hum. Genet.*, 130: 523–632.

Lorenzi, H.A. and R.H. Reeves. 2006. Hippocampal hypocellularity in Ts65Dn mouse originates early in development. *Brain Res.*, 1104: 153–159.

Loth, E. 1912. Beiträge zur Anthropologie der Negerweichteile: Muskelsystem. Studien und Forschungen zur Menschen—und Völkerkunde. Stuttgart.

Loth, E. 1931. Anthropolgie Des Parties Molles (Mucles, Intentions, Visseaux, Nerf Periphriques). Paris: Mianowski—Masson et CIE.

Loth, E. 1950. Anthropological studies of muscles of Uganda Negroes. In *Yearbook of Physical Anthropology.* 1949. (Lasker, G.W. and C.I. Shade Eds.). The Viking Fund, Inc., New York. pp. 220–231.

Loukas, M., R.G. Louis Jr., G. South, E. Alsheik and C. Christopherson. 2006. A case of an accessory brachialis muscle. *Clin. Anat.*, 19: 550–553.

Lu, J., R. Bassel-Duby, A. Hawkins, P. Chang, R. Valdez, H. Wu, L. Gan, J. Shelton, J. Richardson and E. Olson. 2002. Control of Facial Muscle Development by MyoR and Capsulin. *Science*, 298. doi:10.1126/science.1078273.

Macalister, A. 1866. Notes on muscular anomalies in human anatomy. *Proc. R. Irish Acad.*, IX: 444–467.

Macalister, A. 1867. Notes on muscular anomalies in human anatomy. *Proc. R. Irish Acad.*, X: 121–164.

Macalister, A. 1875. Additional observations on muscular anomalies in human anatomy (Third Series), with a catalogue of the principal muscular variations hitherto published. *Trans R. Irish Acad.*, XXV: 1–34.

Madhavi, C. and S.J. Holla. 2003. Anomalous fexor digiti minimi brevis in Guyon's canal. *Clin. Anat.*, 16: 340–343.

Mall, F.P. 1917. Cyclopia in the Human Embryo *Contributions to Embryology Volume VI.* Carnegie Institute of Washington.

Manzano, A., V. Abdala, M.L. Ponssa and M. Soliz. 2013. Ontogeny and tissue differentiation of the pelvic girdle and hind limbs of anurans. *Acta. Zool.*, 94: 420–436.

Marshall, C.R., E.C. Raff and R.A. Raff. 1994. Dollo's law and the death and resurrection of genes. *Proc. Natl. Acad. Sci. U.S.A.*, 91: 12283–12287.

Martin, P. 1990. Tissue patterning in the developing mouse limb. *Int. J. Dev. Biol.*, 34: 323–336.

McGrath, P. 1992. The proboscis in human cyclopia: An anatomical study in two dimensions. *J. Anat.*, 181: 139–149.

McGrath, P. and G.H. Sperber. 1990. Floor of the median orbit in human cyclopia: An anatomical study in three dimensions. *J. Anat.*, 169: 125–138.

Merida-Velasco, J.R., J.F. Rodriguez Vazquez, J.A. Merida Velasco, J. Sobrado Perez and J.J. Collado. 2003. Axillary arch: Potential cause of neurovascular compression syndrome. *Clin. Anat.*, 16: 514–519.

Mieden, G.D. 1982. An anatomical study of three cases of alobar holoprosencephaly. *Teratology,* 26: 123–133.

Moen, D.W., J.K. Werner and E.T. Bersu. 1984. Analysis of gross anatomical variations in human triploidy. *Am. J. Med. Gen.*, 18: 345–56.

Moore, C.S. 2006. Postnatal lethality and cardiac anomalies in the Ts65Dn Down syndrome mouse model. *Mamm. Genome*, 17: 1005–1012.

Moore, C.S. and R.J. Roper. 2007. The power of comparative and developmental studies for mouse models of Down syndrome. *Mamm. Genome*, 18: 431–443.

Morice, E., L.C. Andreae, S.F. Cooke, L. Vanes, E.M. Fisher, V.L. Tybulewicz and T.V. Bliss. 2008. Preservation of long-term memory and synaptic plasticity despite short-term impairments in the Tc1 mouse model of Down syndrome. *Learn. Mem.*, 15: 492–500.

Muntz, L. 1975. Myogenesis in the trunk and leg during development of the tadpole of *Xenopus laevis* (Daudin 1802). *J. Embryol. Exp. Morphol.*, 33: 757–774.

Nadel, L. 2003. Down's Syndrome: a genetic disorder in biobehavioral perspective. *Genes Brain Behav.*, 2: 156–166.

Nakatani, T., S. Tanaka and S. Mizukami. 1998. Bilateral four-headed biceps brachii: The median nerve and brachial artery passing through a tunnel formed by a muscle skip from the accessory head. *Clin. Anat.*, 11: 209–212.

Noden, D.M. 1986. Patterning of avian craniofacial muscles. *Dev. Biol.*, 356: 347–356.

O'Doherty, A., S. Ruf, C. Mulligan, V. Hildreth, M.L. Errington, S. Cooke, A. Sesay, S. Modino, L. Vanes, D. Hernandez, J.M. Linehan, P.T. Sharpe, S. Brandner, T.V. Bliss, D.J. Henderson, D. Nizetic, V.L. Tybulewicz and E.M. Fisher. 2005. An aneuploid mouse strain carrying human chromosome 21 with Down syndrome phenotypes. *Science*, 309: 2033–2037.

Oh, C.S., I.H. Chung and K.S. Koh. 2000. Anatomical study of the accessory head of the flexor pollicis longus and the anterior interosseous nerve in Asians. *Clin. Anat.*, 13: 434–438.

Olson, L.E. and S. Mohan. 2011. Bone density phenotypes in mice aneuploid for the Down syndrome critical region. *Am. J. Med. Genet. A*, 155: 2436–2445.

Olson, L.E., J.T. Richtsmeier, J. Leszl and R.H. Reeves. 2004. A Chromosome 21 critical region does not cause specific Down syndrome phenotypes. *Science*, 306: 687–690.

Olson, L.E., R.J. Roper, E.A. Sengstaken, E.A. Peterson, V. Aquino, Z. Galdzicki, R. Siarey, M. Pletnikov, T.H. Moran and R.H. Reeves. 2007. Trisomy for the Down syndrome "critical region" is necessary but not sufficient for brain pheotypes of trisomic mice. *Hum. Mol. Genet.*, 16: 774–782.

Opitz, J.M. and E.F. Gilbert-Barness. 1990. Reflections on the pathogenesis of Down syndrome. *Am. J. Med. Gen. Supplement*, 7: 38–51.

Opitz, J.M. 1985. The developmental field concept. *Am. J. Med. Gen.*, 11: 1–11.

Ovalle, W.K. 1999. Morphological study of two human facial muscles: Orbicularis oculi and corrugator supercilii. *Clin. Anat.*, 11: 1–11.

Owen, R. 1849. *On the nature of limbs.* John Van Voorst, London.

Pai, M.M., S.R. Nayak, A. Krishnamurthy, R. Vadgaonkar, L.V. Prabhu, A.V. Ranade, J.P. Janardhan and R. Rai. 2008. The accessory heads of flexor pollicis longus and flexor digitorum profundus: Incidence and morphology. *Clin. Anat.*, 21: 252–258.

Paisant, S., R. Sambasivan, B. Gayraud-Morel, R.G. Kelly and S. Tajbakhsh. 2009. Distinct regulatory cascades govern extraocular and pharyngeal arch muscle progenitor cell fates. *Dev. Cell*, 5: 810–821.

Penhall, B., G. Townsend, S. Tomo and K. Nakajima. 1998. The pterygoideus proprius muscle revisited. *Clin. Anat.*, 11: 332–337.

Perelman, P., W.E. Johnson, C. Roos, H.N. Seuánez, J.E. Horvath, M.A. Moreira, B. Kessing, J. Pontius, M. Roelke, Y. Rumpler, M.P.P. Schneider, A. Silva, S.J. O'Brien and J. Pecon-Slattery. 2011. A molecular phylogeny of living primates. *PLoS Genetics*, 7:e1001342.

Pessa, J.E., V.P. Zadoo, P.A. Garza, E.K. Adrian Jr., A.I. DeWitt and J.R. Garza. 1998. Double or bifid zygomaticus major muscle: Anatomy incidence and clinical correlation. *Clin. Anat.*, 11: 310–313.

Pettersen, J.C. 1979. Anatomical studies of a boy trisomic for the distal portion of 13q. *Am. J. Med. Gen.*, 4: 383–400.

Pettersen, J.C., G.G. Koltis and M.G. White. 1979. An examination of the spectrum of anatomic defects and variations found in eight cases of Trisomy 13. *Am. J. Med. Gen.*, 3: 183–210.

Plock, J., C. Contaldo and M. von Ludin-Ghausen. 2005. Levator ppalpebrae superioris muscle in human fetuses: Anatomical findings and their clinical relevance. *Clin. Anat.*, 18: 473–480.

Ponssa, M.L., J. Goldberg and V. Abdala. 2010. Sesamoids in anurans: New data, old issues. *Anat. Rec.*, 293: 1646–1668.

Presch, W. 1975. The evolution of limb reduction in the teiid lizard genus *Bachia. Bull. Sci. Calif. Acad. Sci.*, 74: 113–121.

Prunotto, C., T. Crepaldi, P.E. Forni, A. Ieraci, R.G. Kelly, S. Tajbakhsh, M. Buckingham and C. Ponzetto. 2004. Analysis of *Mlc-lacZ* Met mutants highlights the essential function of Met for migratory precursors of hypaxial muscles and reveals a role for Met in the development of hyoid arch-derived facial muscles. *Dev. Dyn.*, 231: 582–591.

Ragoowansi, R., A. Adeniran and A.L. Moss. 2002. Anomalous muscle of the wrist. *Clin. Anat.*, 15: 363–365.

Ramirez-Castro, J.L. and E.T. Bersu. 1978. Anatomical analysis of the developmental effects of aneuploidy in Man—The 18 trisomy syndrome: II. Anomalies of the upper and lower limbs. *Am. J. Med. Gen.*, 2: 285–306.

Reeves, R.H., N.G. Irving, T.H. Moran, A. Wohn, C. Kitt, S.S. Sisodia, C. Schmidt, R.T. Bronson and M.T. Davisson. 1995. A mouse model for Down syndrome exhibits learning and behaviour deficits. *Nat. Genet.* 11: 177–184.

Reinholdt, L.G., Y. Ding, G.L. Gilbert, A. Czechanski, J.P. Solzak, R.J. Roper, M.T. Johnson, L.R. Donahue, C. Lutz and M.T. Davisson. 2011. Molecular characterization of the translocation breakpoints in the Down syndrome mouse model Ts65Dn. *Mamm. Genome*, 22: 685–691.

Reynolds, L.E., A.R. Watson, M. Baker, T.A. Jones, G. D'Amico, S.D. Robinson, C. Joffre, S. Garrido-Urbani, J.C. Rodriguez-Manzaneque, E. Martino-Echarri, M. Aurrand-Lions, D. Sheer, F. Dagna-Bricarelli, D. Nizetic, C.J. McCabe, A.S. Turnell, S. Kermorgant, B.A. Imhof, R. Adams, E.M. Fisher, V.L. Tybulewicz, I.R. Hart and K.M. Hodivala-Dilke. 2010. Tumour angiogenesis is reduced in the Tc1 mouse model of Down's syndrome. *Nature*, 465: 813–817.

Richtsmeier, J.T., L.L. Baxter and R.H. Reeves. 2000. Parallels of craniofacial maldevelopment in Down syndrome and Ts65Dn mice. *Dev. Dyn.*, 217: 137–145.

Rodríguez-Niedenführ, M., T. Vázquez, P. Golanó, I. Parkin and J.R. Sañudo. 2002. Extensor digitorum brevis manus: Anatomical radiological and clinical relevance. A review. *Clin. Anat.*, 15: 286–292.

Rolian, C. 2009. Integration and evolvability in primate hands and feet. *Evol. Biol.*, 36: 100–117.

Rolian, C., D.E. Lieberman and B. Hallgrímsson. 2010. The coevolution of human hands and feet. *Evol.; Intern. J. Org. Evol.*, 64: 1558–1568.

Romer, A.S. 1933. *The Osteology of the Reptiles.* The University of Chicago Press, Chicago, 772pp.

Roper, R.J. and R.H. Reeves. 2006. Understanding the basis for Down syndrome phenotypes. *PLoS Genet.*, 2: e50.

Roper, R.J., L.L. Baxter, N.G. Saran, D.K. Klinedinst, P.A. Beachy and R.H. Reeves. 2006a. Defective cerebellar response to mitogenic Hedgehog signaling in Down syndrome mice. *Proc. Natl. Acad. Sci. U.S.A.*, 103: 1452–1456.

Roper, R.J., H.K. St. John, J. Philip, A. Lawler and R.H. Reeves. 2006b. Perinatal loss of Ts65Dn "Down syndrome" mice. *Genetics*, 172: 437–443.

Roper, R.J., J.F. VanHorn, C.C. Cain and R.H. Reeves. 2009. A neural crest deficit in Down syndrome mice is associated with deficient mitotic response to Sonic Hedgehog. *Mech. Dev.*, 126: 212–219.

Rosenheimer, J.L., J. Loewy and S. Lozanoff. 2000. Levator claviculae muscle discovered during physical examination for cervical lymphadenopathy. *Clin. Anat.*, 13: 298–301.

Rothermel, B., R.B. Verga, J. Yang, H. Wu, R. Bassel-Duby and R.S. Williams. 2000. A protein encoded within the Down syndrome critical region is enriched in striated muscles and inhibits calcineurin signaling. *J. Biol. Chem.*, 275: 8719–8725.

Ruge, G. 1887. *Untersuchungen über die Gasichtmuskulatur der Primaten.* Leipzig.

Ruvinsky, I. and J.J. Gibson-Brown. 2000. Genetic and developmental bases of serial homology in vertebrate limb evolution. *Development*, 127: 5233–44.

Sambasivan, R., S. Kuratani and S. Tajbakhsh. 2011. An eye on the head: The development and evolution of craniofacial muscles. *Development*, 2415: 2401–2415.

Sanna, B., E.B. Brandt, R.A. Kaiser, P. Pfluger, S.A. Witt, T.R. Kimball, E. van Rooij, L.J. De Windt, M.E. Rothenberg, M.H. Tschop, S.C. Benoit and J.D. Molkentin. 2006. Modulatory calcineurin-interacting proteins 1 and 2 function as calcineurin facilitators *in vivo*. *PNAS*, 103: 7327–7332.

Schmidt, M. and M.S. Fischer. 2009. Morphological integration in mammalian limb proportions: dissociation between function and development. *Evol.; Internat. J. Org. Evol.*, 63: 749–766.

Schowing, J. 1968. Influence inductrice de l'encephale embryonnaire sur le developpement du crene chez le poulet. *J. Embryol. Exp. Morphol.*, 1: 23–32.

Schön Ybarra, M.A. and B. Bauer. 2001. Medial portion of m.temporalis and its potential involvement in facial pain. *Clin. Anat.*, 14: 25–30.

Scott-Conner, C.E.H. and A.S. Al-Jurf. 2002. The sternalis muscle. *Clin. Anat.*, 15: 67–69.

Shapiro, B.L. 1983. Down syndrome—A disruption of homeostasis. *Am. J. Med. Gen.*, 269: 241–269.

Shou, S., V. Scott, C. Reed, R. Hitzemann and H.S. Stadler. 2005. Transcriptome analysis of the murine forelimb and hindlimb autopod. *Dev. Dyn.*, 234: 74–89.

Shubin, N.H. and P. Alberch. 1986. A morphogenetic approach to the origin and basic organization of the tetrapod limb. *Evol. Biol.*, 20: 319–387.

Shubin, N., C. Tabin and S. Carroll. 1997. Fossils, genes and the evolution of animal limbs. *Nature*, 388: 639–48.

Shwartz, Y., Z. Farkas, T. Stern, A. Aszódi and E. Zelzer. 2012. Muscle contraction controls skeletal morphogenesis through regulation of chondrocyte convergent extension. *Dev. Biol.* 370: 154–163.

Siarey, R.J., J. Stoll, S.I. Rapoport and Z. Gladzicki. 1997. Altered long term potentiation in the young and old Ts65Dn mouse, a model for Down syndrome. *Neuropharmacology*, 36: 1549–1554.

Siegal, M.L. and A. Bergman. 2002. Waddington's canalization revisited: Developmental stability and evolution. *Proc. Nat. Ac. Sci. U.S.A.*, 99: 10528–32.

Situ, D., C.W. Reifel, R. Smith, G.W. Lyons, R. Temkin, C. Harper-Little and S.C. Pang. 2002. Investigation of a cyclopic, human, term fetus by use of Magnetic Resonance Imaging (MRI). *J. Anat.*, 200: 431–438.

Smith, S. and B. Boulgakow. 1926. A case of cyclopia. *J. Anat.*, 61: 105–111.

Snell, R.S., 1995. *Clinical Anatomy for Medical Students 5th EDT.* Boston: Little, Brown & Co. Inc.

Solzak, J.P., Y. Liang, F.C. Zhou and R.J. Roper. 2013. Commonality in Down and fetal alcohol syndromes. *Birth Defects Res. A*, 97: 187–197.

Sperber, G.H., L.H. Honore and G.A. Machin. 1989. Microscopic study of holoprosencephalic facial anomalies in Trisomy 13 fetuses. *Am. J. Med. Gen.*, 32: 443–451.

Stark, H.H., T.A. Otter, J.H. Boyes and T.A. Rickard. 1979. Atavistic contrahentes digitorum and associated muscle abnormalities of the hand: A cause of symptoms. *J. Bone Jt. Surg.*, 61A: 286–289.

Starbuck, J.M., T.M. Cole III, R.H. Reeves and J.T. Richtsmeier. 2013. Trisomy 21 and facial developmental instability. *Am. J. Phys. Anthro.*, 151: 49–57.

Stempfle, N., Y. Huten, C. Fredouille, H. Brisse and C. Nessmann. 1999. Skeletal abnormalities in fetuses with Down's syndrome: A radiographic post-mortem study. *Pediatric Radiology*, 29: 682–688.

Stoll, C., Y. Alembik, B. Dott and M.P. Roth. 1998. Study of Down syndrome in 238,942 consecutive births. *Ann. Genet.*, 41: 44–51.

Testut, L. 1888. *Les anomalies musculaires chez l'homme, expliquées par l'anatomie comparée: leur importance en anthropologie.* Paris: Masson.

Thakur, S., R. Singh, M. Pradhan and R. Phadke. 2004. Spectrum of holoprosencephaly. *Indian Journal of Pediatrics.* 71: 593–597.

Tiengo, C., V. Macchi, C. Stecco, F. Bassetto and R. De Caro. 2006. Epifascial accessory palmaris longus muscle. *Clin. Anat.*, 19: 554–557.

Tonkin, M.A. and N.W. Bulstrode. 2007. Bilhaut-Cloquet procedure for Wassel types III, IV and VII thumb duplication. *J. Hand Surg. Eur.*, 32: 684–693.

Tubbs, R.S., E.G. Salter and W.J. Oakes. 2005. Contrahentes digitorum muscle. *Clin. Anat.*, 18: 606–608.

Tubbs, R.S., E.G. Salter and W.J. Oakes. 2006a. Dissection of a rare accessory muscle of the leg: The tensor fasciae suralis muscle. *Clin. Anat.*, 19: 571–572.

Tubbs, R.S., E.G. Salter and W.J. Oakes. 2006b. The psoas quartus muscle. *Clin. Anat.*, 19: 678–680.

Tubbs, R.S., E.G. Salter and W.J. Oakes. 2006c. Triceps brachii muscle demonstrating a fourth head. *Clin. Anat.*, 19: 657–660.

Turgut, H.B., T. Peker, N. Gülekon, A. Anil and K. Mustafa. 2005. Axillopectoral muscle (Langer's muscle). *Clin. Anat.*, 18: 220–223.

Tyler, M. 1983. Development of the frontal bone and cranial meninges in the embryonic chick : An experimental study of tissue interactions. *Anat. Rec.*, 70: 61–70.

Tzahor, E. 2009. Heart and craniofacial muscle development: A new developmental theme of distinct myogenic fields. *Dev. Biol.*, 327: 273–279.

Tzahor, E. and S.M. Evans. 2011. Pharyngeal mesoderm development during embryogenesis: Implications for both heart and head myogenesis. *Cardiovasc. Res.*, 2: 196–202.

Urban, B. and E.T. Bersu. 1987. Chromosome 18 aneuploidy: Anatomical variations observed in cases of full and mosaic Trisomy 18 and a case of deletion of the short arm of Chromosome 18. *Am. J. Med. Gen.*, 27: 425–34.

Van Cleve, S.N. and W.I. Cohen. 2006. Part I: clinical practice guidelines for children with Down syndrome from birth to 12 years. *J. Pediatr. Health Care*, 20: 47–54.

Van Cleve, S.N., S. Cannon and W.I. Cohen. 2006. Part II: Clinical Practice Guidelines for adolescents and young adults with Down syndrome: 12 to 21 Years. *J. Pediatr. Health Care*, 20: 198–205.

Wade, N. 2014. *Troublesome inheritance, genes, race and human history.* New York: The Penguin Press.

Wagner, G.P. and C.H. Chiu. 2001. The tetrapod limb: a hypothesis on its origin. *J. Exp. Zool. B (Mol. Dev. Evol.)* 291: 226–240.

Wagner, G.P. and H.C.E. Larsson. 2007. Fins and limbs in the study of the evolutionary novelties. In *Fins into Limbs. Evolution, Development, and Transformation* (Hall, B.K. Ed.). The University of Chicago Press, Chicago and London. pp. 49–61.

Wahba, M.Y., G.D. Singh and S. Lozanoff. 1998. An anomalous accessory flexor digiti minimi profundus muscle: A case study. *Clin. Anat.*, 11: 55–59.

Waters, P.M. and D.S. Bae. 2012. *Pediatric Hand and Upper Limb Surgery: A Practical Guide.* Lippincott Williams and Wilkins, London.

Watt, A.J. and K.C. Chung. 2009. Duplication. *Hand Clin.* 25: 215–227.

Weatherbee, S.D. and L.A. Niswander. 2007. Mechanisms of chondrogenesis and osteogenesis in limbs. *Fins into Limbs. Evolution, Development, and Transformation* (Hall, B.K. Ed.). The University of Chicago Press, Chicago and London. pp. 93–102.

West-Eberhard, M.J. 2003. Developmental plasticity and evolution. Oxford University Press, Oxford.

Wiens, J.J. 2011. Re-evolution of lost mandibular teeth in frogs after more than 200 million years, and re-evaluating Dollo's law. *Evolution*, 65: 1283–1296.

Wood, B. and N.L. Lonergan. 2008. The hominin fossil record: taxa, grades and clades. *J. Anat.*, 212: 354–376.

Wood, J. 1864. On some varieties in human myology. *Proc. R. Soc. (Lond.)*, 13: 299–303.

Wood, J. 1865. Additional varieties in human myology. *Proc. R. Soc. (Lond.)*, 14: 379–393.

Wood, J. 1866. Variations in human myology observed during the winter session of Winter Session of 1865–66 at King's College, London. *Proc. R. Soc. (Lond.)*, 15: 229–244.

Wood, J. 1867a. Variations in human myology observed during the winter session of Winter Session of 1866–67 at King's College, London. *Proc. R. Soc. (Lond.)*, 518–545.

Wood, J. 1867b. On human muscular variations and their relation to comparative anatomy. *J. Anat. Physiol.*, I: 44–59.

Wood, J. 1868. Additional varieties in human myology observed during the winter session of Winter Session of 1867–68 at King's College, London. *Proc. R. Soc. (Lond.)*, 16: 483–525.

Wood, J. 1869. On a group of varieties of the muscles of the human neck, shoulder, and chest, with their transitional forms and homologies in mammalia. *Proc. R. Soc. (Lond.),* 18: 1–3.

Wood, J. 1870. On a group of varieties of the muscles of the human neck, shoulder, and chest, with their transitional forms and homologies in mammalia. *Tran. R. Soc. (Lond.),* 160: 83–116.

Young, R.L., V. Caputo, M. Giovannotti, T. Kohlsdorf, A.O. Vargas, G.E. May and G.P. Wagner. 2009. Evolution of digit identity in the three-toed Italian skink *Chalcides chalcides*: a new case of digit identity frame shift. *Evol. Dev.* 11: 647–658.

Young, N.M. and B. Hallgrímsson. 2005. Serial homology and the evolution of mammalian limb covariation structure. *Evol.; Intern. J. Org. Evol.,* 59: 2691–704.

Young, N.M., G.P. Wagner and B. Hallgrímsson. 2010. Development and the evolvability of human limbs. *Proc. Nat. Acad. Sci. U.S.A.,* 107: 3400–3405.

Yu, T., Z. Li, Z. Jia, S.J. Clapcote, C. Liu, S. Li, S. Asrar, A. Pao, R. Chen, N. Fan, S. Carattini-Rivera, A.R. Bechard, S. Spring, R.M. Henkelman, G. Stoica, S. Matsui, N.J. Nowak, J.C. Roder, C. Chen, A. Bradley and Y.E. Yu. 2010. A mouse model of Down syndrome trisomic for all human chromosome 21 syntenic regions. *Hum. Mol. Genet.,* 19: 2780–2791.

Dissection Photographs of Trisomy 18 Human Cyclopia Fetus

Cyclopic eye

Left
Superior ←→ Inferior
Right

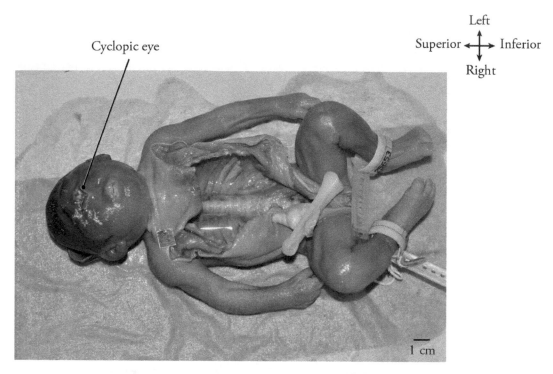

Figure A.1. Anterior view of the body showing previously dissected areas.

Right
Superior ←→ Inferior
Left

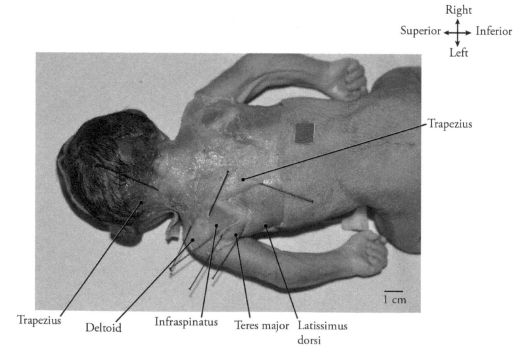

Trapezius

Trapezius   Deltoid   Infraspinatus   Teres major   Latissimus dorsi

Figure A.2. Posterior view of body with superficial back dissection.

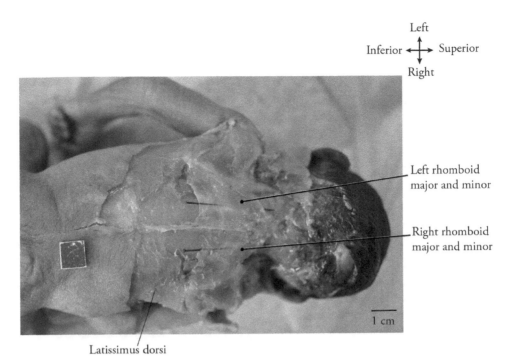

Figure A.3. Posterior view of body with deep back dissection.

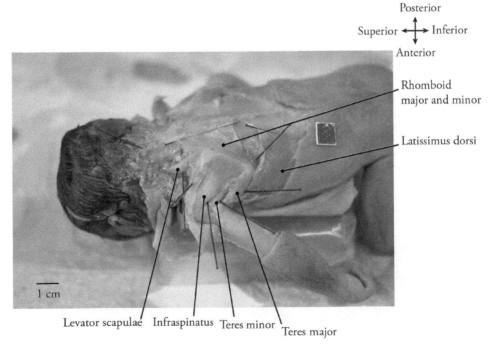

Figure A.4. Left shoulder dissection.

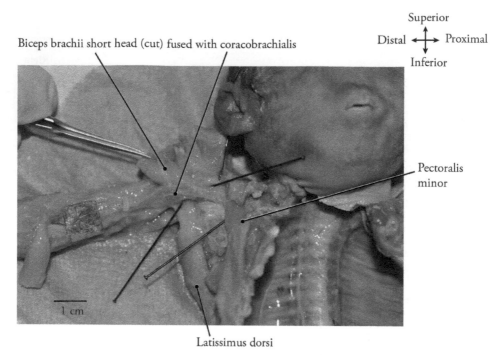

Figure A.5. Right anterior deep axillary dissection.

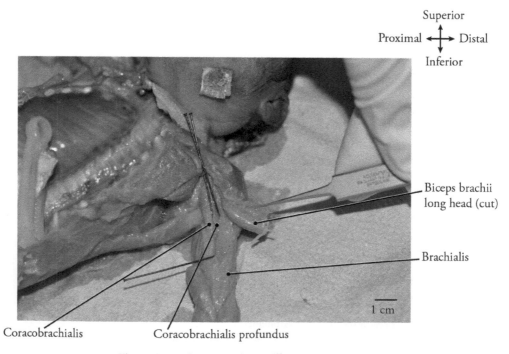

Figure A.6. Left anterior deep axillary dissection.

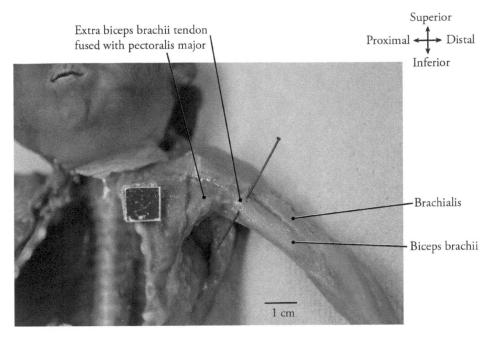

Figure A.7. Left anterior superficial axillary dissection.

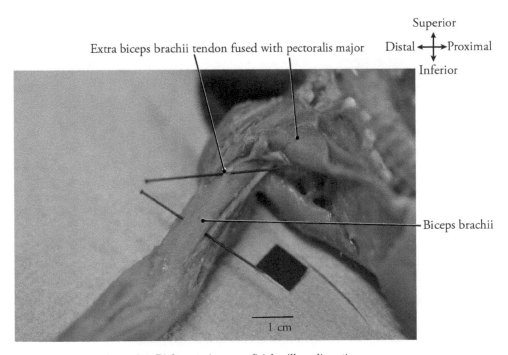

Figure A.8. Right anterior superficial axillary dissection.

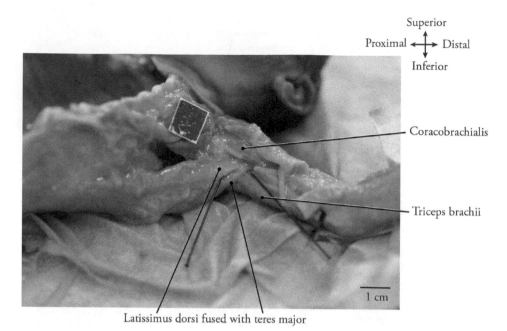

Superior

Proximal ←→ Distal

Inferior

Coracobrachialis

Triceps brachii

1 cm

Latissimus dorsi fused with teres major

Figure A.9. Inferior view of left axilla.

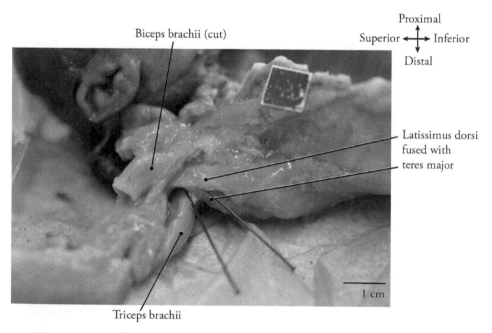

Proximal

Superior ←→ Inferior

Distal

Biceps brachii (cut)

Latissimus dorsi fused with teres major

1 cm

Triceps brachii

Figure A.10. Inferior view of right axilla.

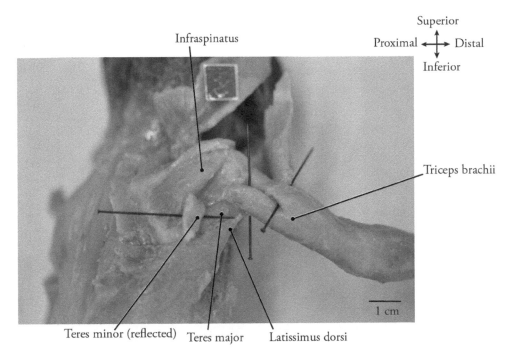

Figure A.11. Posterior right shoulder dissection.

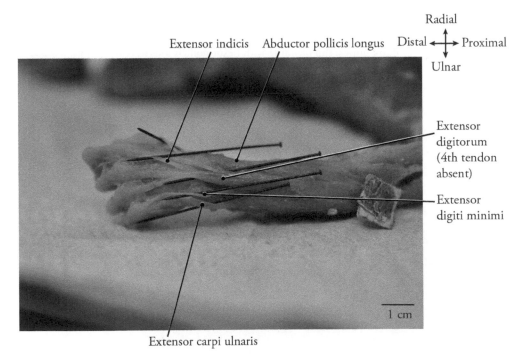

Figure A.12. Extensors of the left forearm.

Flexor carpi radialis   Brachioradialis

Radial
Distal ←→ Proximal
Ulnar

Biceps
brachii (cut)

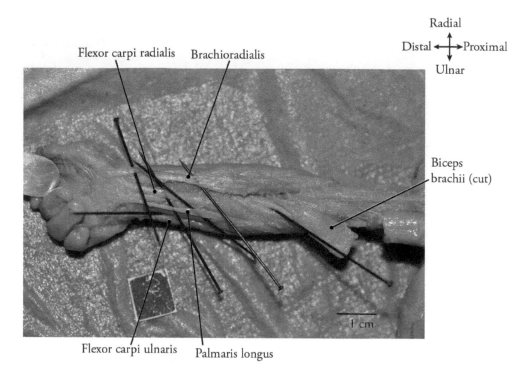

1 cm

Flexor carpi ulnaris   Palmaris longus

Figure A.13. Flexors of the right forearm.

Extra muscle slip blending with flexor pollicis longus (cut)

Ulnar
Distal ←→ Proximal
Radial

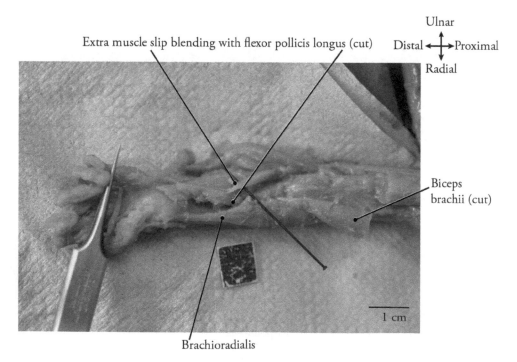

Biceps
brachii (cut)

1 cm

Brachioradialis

Figure A.14. Deep view of the left forearm flexors.

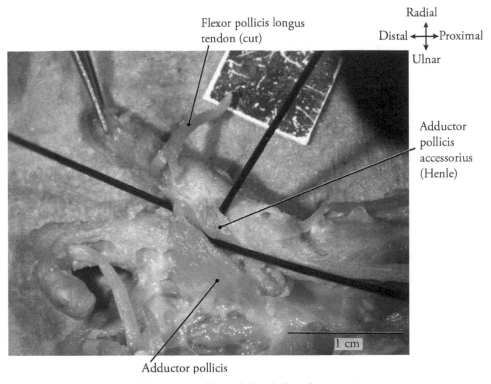

Flexor pollicis longus
tendon (cut)

Radial
Distal ←→ Proximal
Ulnar

Adductor
pollicis
accessorius
(Henle)

1 cm

Adductor pollicis

Figure A.15. Palmar view of the right hand, deep thenar region.

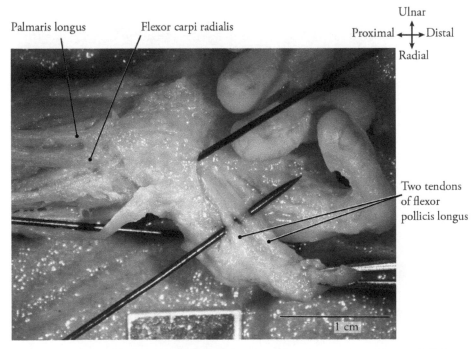

Palmaris longus   Flexor carpi radialis

Ulnar
Proximal ←→ Distal
Radial

Two tendons
of flexor
pollicis longus

1 cm

Figure A.16. Palmar view of the right hand, superficial thenar area. Note the complete absence of all thenar muscles.

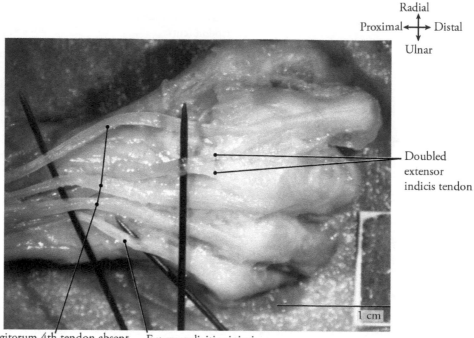

Radial
Proximal ←→ Distal
Ulnar

Doubled
extensor
indicis tendon

1 cm

Extensor digitorum 4th tendon absent   Extensor digiti minimi

Figure A.17. Extensors of the right hand.

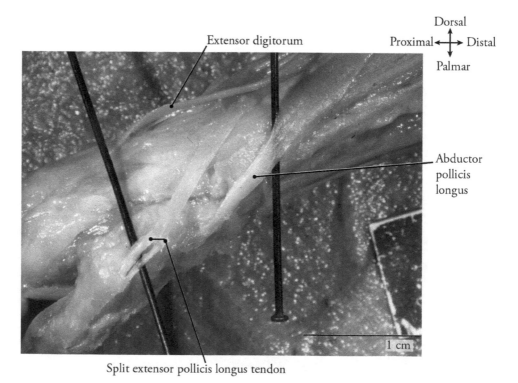

Extensor digitorum

Dorsal
Proximal ←→ Distal
Palmar

Abductor
pollicis
longus

1 cm

Split extensor pollicis longus tendon

Figure A.18. Radial view of the right thumb.

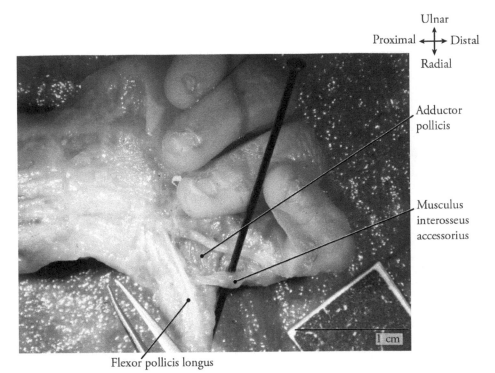

Figure A.19. Palmar view of the right thenar region.

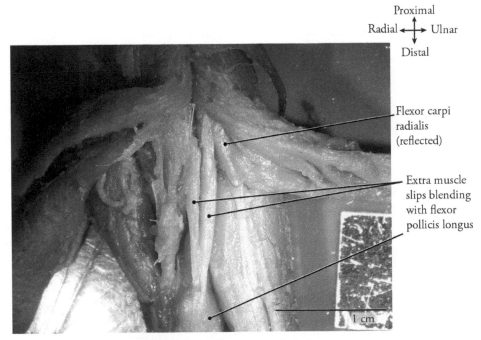

Figure A.20. Deep flexor view of the right forearm.

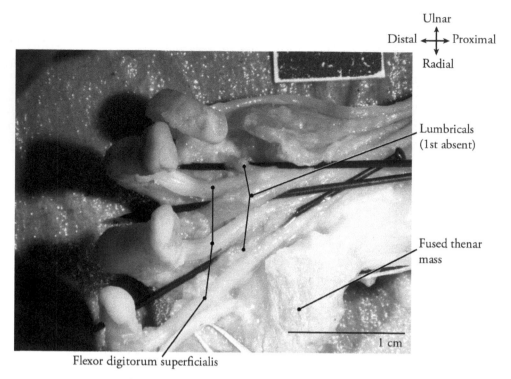

Figure A.21. Palmar dissection of the left hand.

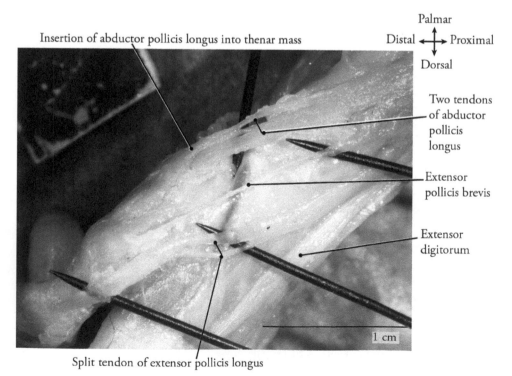

Figure A.22. Radial view of the left thumb.

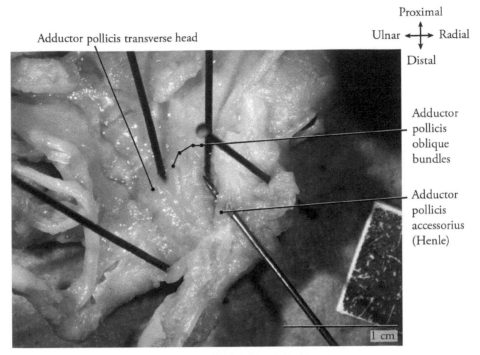

Figure A.23. Deep palmar dissection of the left hand.

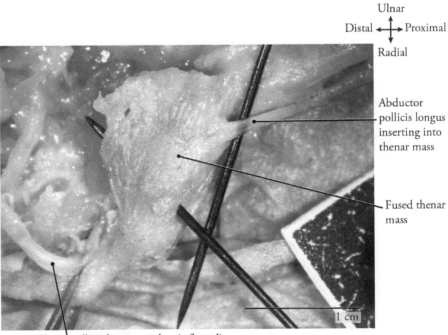

Figure A.24. Thenar region of the left hand.

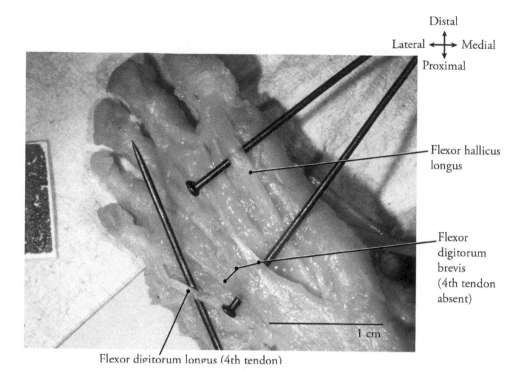

Distal

Lateral ←→ Medial

Proximal

Flexor hallicus longus

Flexor digitorum brevis (4th tendon absent)

1 cm

Flexor digitorum longus (4th tendon)

Figure A.25. Superficial plantar view of the right foot.

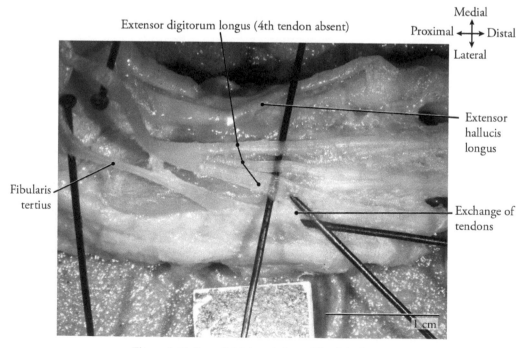

Extensor digitorum longus (4th tendon absent)

Medial

Proximal ←→ Distal

Lateral

Extensor hallucis longus

Fibularis tertius

Exchange of tendons

1 cm

Figure A.26. Superficial dorsal view of the right foot.

Flexor digitorum brevis (4th tendon absent)

Proximal

Lateral ← → Medial

Distal

Flexor hallucis brevis

Flexor hallucis longus

1 cm

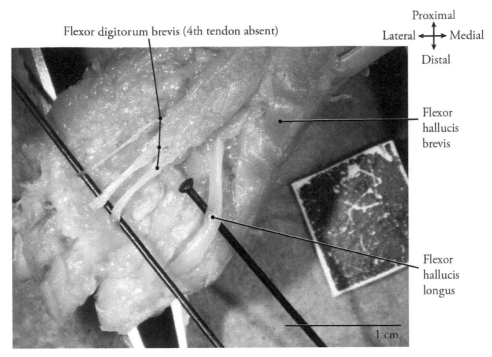

Figure A.27. Superficial plantar view of the left foot.

Medial

Distal ← → Proximal

Lateral

Fibularis tertius

Extensor digitorum longus

1 cm

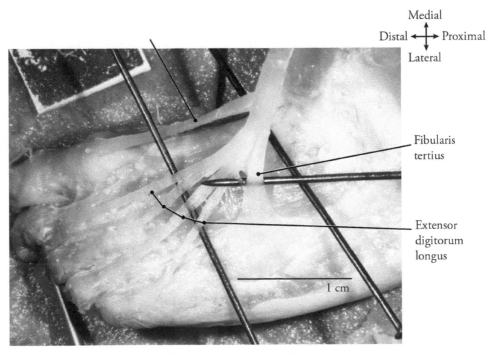

Figure A.28. Superficial dorsal view of the left foot.

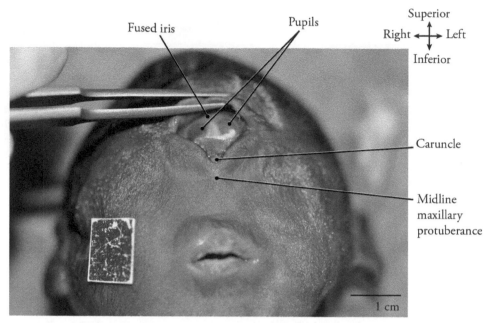

Figure A.29. Anterior view of the face and fused eyes.

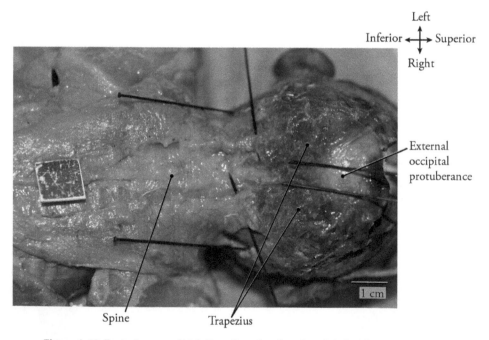

Figure A.30. Posterior superficial dissection of neck and occipital region.

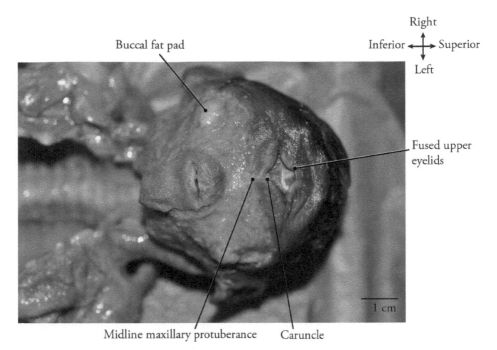

Figure A.31. Anterior superficial facial dissection.

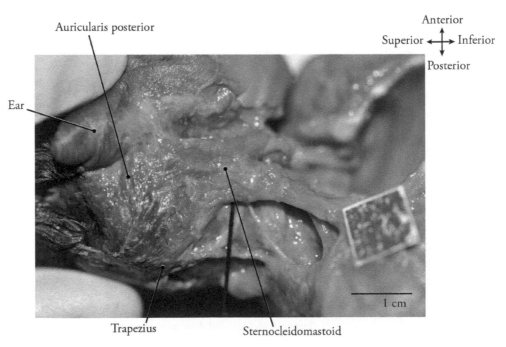

Figure A.32. Right neck dissection.

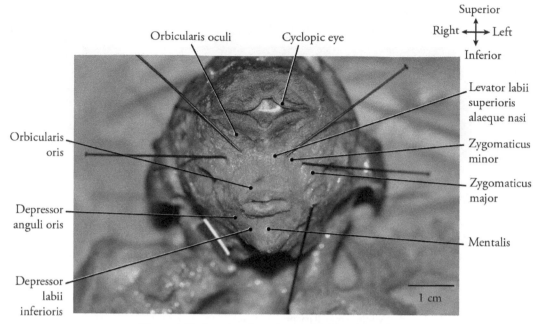

Figure A.33. Anterior view of superficial facial muscles.

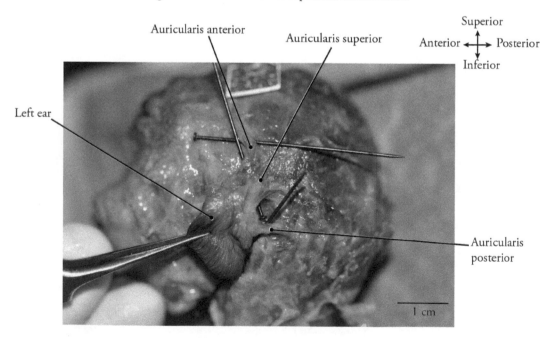

Figure A.34. Left head dissection.

Sternocleidomastoid   Auricularis posterior   Auricularis superior

Superior
Posterior ←→ Anterior
Inferior

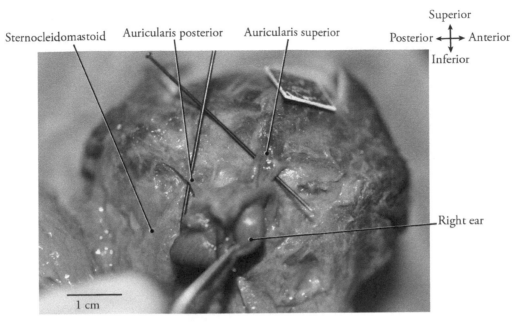

Right ear

1 cm

Figure A.35. Right head dissection.

Zygomaticus major   Orbicularis oculi

Anterior
Inferior ←→ Superior
Posterior

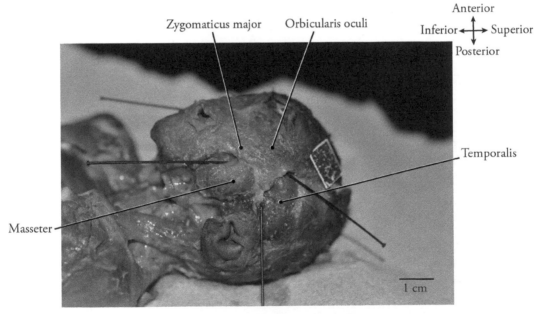

Temporalis

Masseter

1 cm

Figure A.36. Left masticatory muscles.

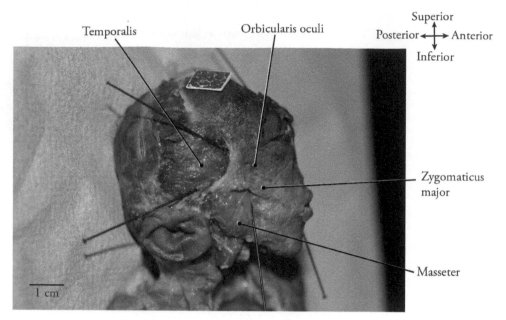

Figure A.37. Right masticatory muscles.

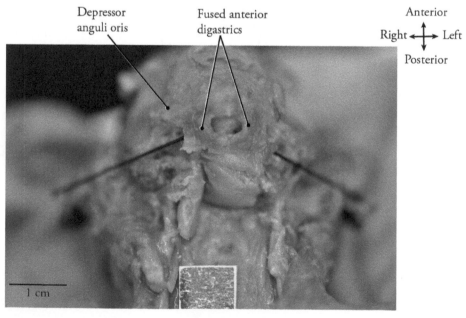

Figure A.38. Inferior view of hyoid region.

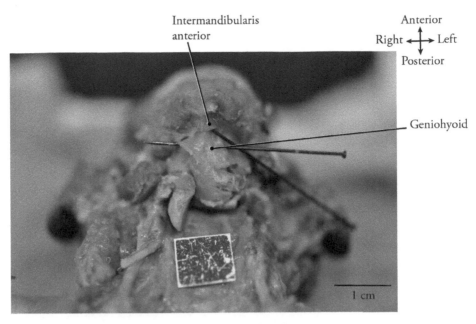

Figure A.39. Inferior view of deep hyoid region.

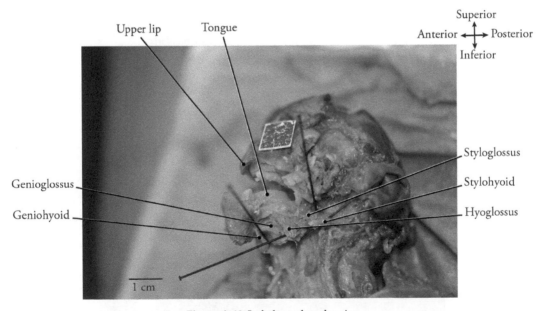

Figure A.40. Left deep glossal region.

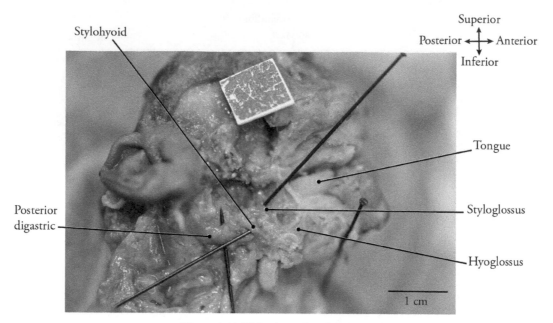

Figure A.41. Right deep glossal region.

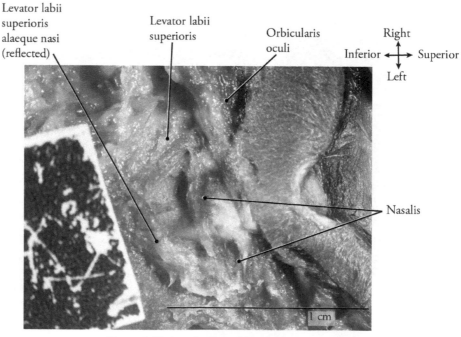

Figure A.42. Anterior face, deep view.

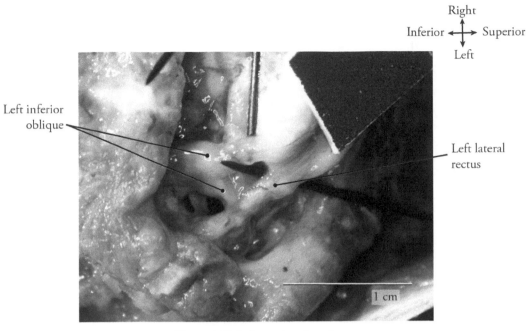

Figure A.43. Left anterior view of eye.

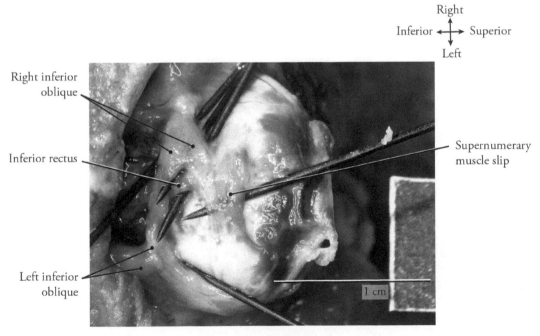

Figure A.44. Inferior view of eye.

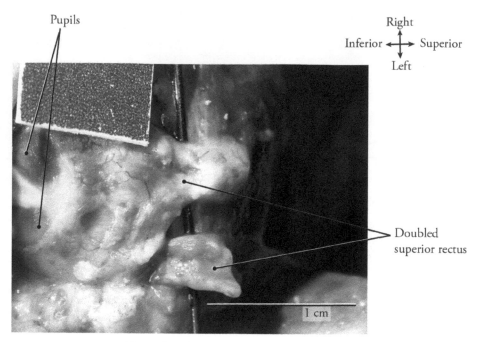

Figure A.45. Superior view of eye.

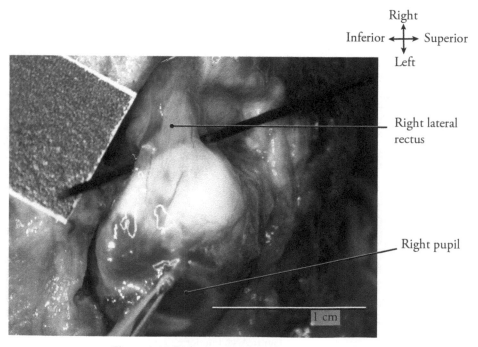

Figure A.46. Right anterior view of eye.

<u>Appendix B</u>

3-D Renders of Trisomy 18 Human Cyclopia Fetus CT Scan Data

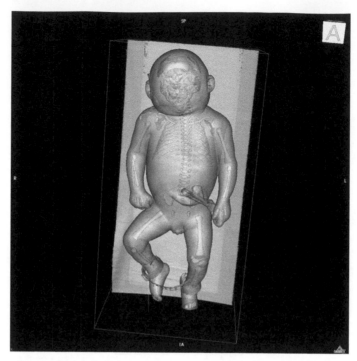

Figure B.1. Anterior view surface render. Transparent skin is shown over skeleton. Render was completed using Osirix DICOM viewer.

Color image of this figure appears in the color plate section at the end of the book.

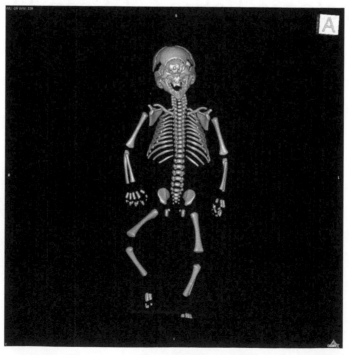

Figure B.2. Anterior view volume render of cyclopia skeleton. Render was completed using Osirix DICOM viewer.

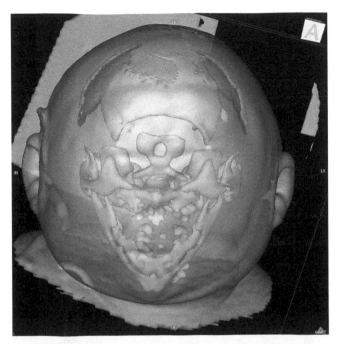

Figure B.3. Anterior view surface render of the cyclopia head. Transparant skin is shown over skull. Render was completed using Osirix DICOM viewer.

Color image of this figure appears in the color plate section at the end of the book.

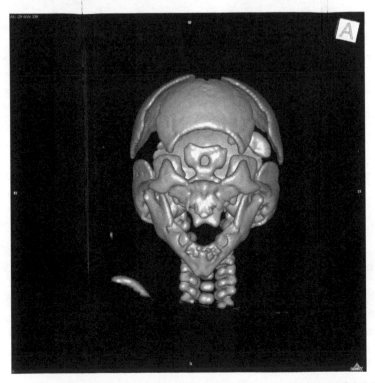

Figure B.4. Anterior view volume render of cyclopia skull. Render was completed using Osirix DICOM viewer.

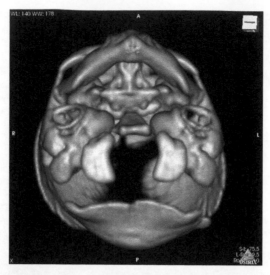

Figure B.5. Inferior view volume render of cyclopia skull base. Render was completed using Osirix DICOM viewer.

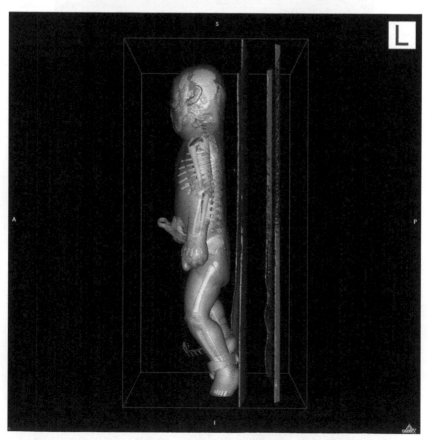

Figure B.6. Left lateral view surface render. Transparent skin is shown over skeleton. Render was completed using Osirix DICOM viewer.

Color image of this figure appears in the color plate section at the end of the book.

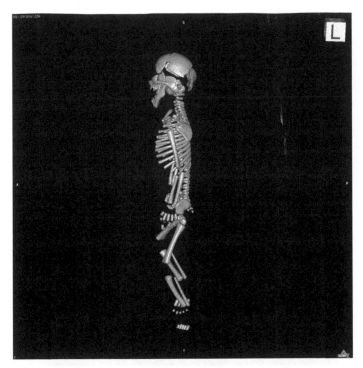

Figure B.7. Left lateral view volume render of cyclopia skeleton. Render was completed using Osirix DICOM viewer.

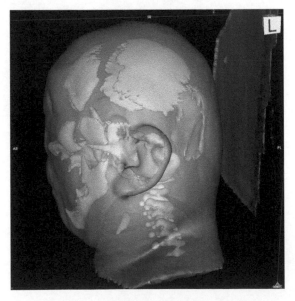

Figure B.8. Left lateral view surface render of cyclopia head. Transparent skin is shown over skull. Render was completed using Osirix DICOM viewer.

Color image of this figure appears in the color plate section at the end of the book.

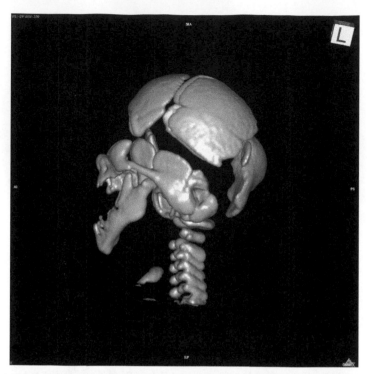

Figure B.9. Left lateral view volume render of cyclopia skull. Render was completed using Osirix DICOM viewer.

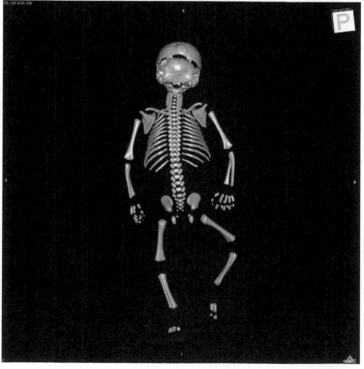

Figure B.10. Posterior view volume render of cyclopia skeleton. Render was completed using Osirix DICOM viewer.

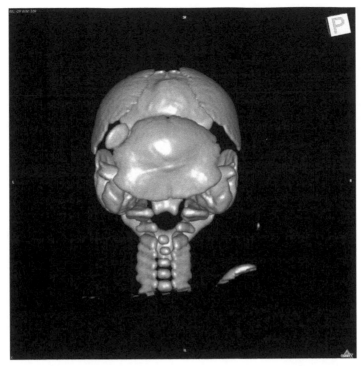

Figure B.11. Posterior view volume render of cyclopia skull. Render was completed using Osirix DICOM viewer.

Index

About the Authors

Christopher M. Smith received his M.A. from the Johns Hopkins Univ. School of Medicine in Medical and Biological Illustration and is currently at Howard University College of Medicine. He combines biomedical research with traditional and digital visualization techniques, and received several awards (e.g., Assoc. Medical Illustrators, Vesalius Trust), including the Vesalius award for the best young medical illustrator and researcher within North America (US and Canada). His current research focuses on the evolutionary and developmental anatomy of humans with and without birth defects.

Janine M. Ziermann is an Instructor and Researcher at the Howard University College of Medicine. She has published several papers on cranial muscle development and evolution in several vertebrate groups, including humans. Her current research focuses on evolution and development of cranial and neck muscles, changes in developmental timing (heterochrony), and evolutionary mechanisms that influence developmental patterns.

M. Ashraf Aziz is Professor at Howard University College of Medicine. He is an evolutionary morphologist whose main interest lies in the comparative myology of human and nonhuman primates. In addition to his numerous independent scientific papers, he has co-authored books with Rui Diogo and his colleagues.

Julia L. Molnar is Visiting Assistant Professor of Comparative Anatomy at Coastal Carolina University. Her research focuses on evolutionary changes in musculoskeletal anatomy and their effects on function, particularly locomotion. She has published on the evolution of locomotion in early tetrapods and crocodylomorphs and the use of virtual models in paleontology, and she is also an awarded medical illustrator.

Marjorie C. Gondré-Lewis is Associate Professor at Howard University College of Medicine. Her research focuses on genetic errors and environmental insults that result in disrupted developmental programs. She has published many research articles and book chapters on cellular mechanisms involved in normal development, on protracted homeostatic, structural and molecular changes induced by perinatal exposure to drugs of abuse, toxins, and stress, and on genetic defects and resulting metabolic and endophenotypes of Niemann-Pick disease, Smith-Lemli-Opitz syndrome, and Edwards syndrome.

Corinne Sandone is a MA, CMI, Certified Medical Illustrator and the director of the graduate program in medical and biological illustration at Johns Hopkins University School of Medicine. She has taught at Hopkins since 1990 and is the illustrator and co-author of several surgical atlases. She is President of the Association of Medical Illustrators (AMI) 2014–2015.

Edward T. Bersu is Professor Emeritus at the University of Wisconsin School of Medicine and Public Health and an Honorary Associate at the School of Engineering at UW-

Madison. His first twenty years of research involved human gross anatomical studies of individuals with trisomy, with an emphasis on trisomy 21 (Down syndrome), and other defined congenital malformation syndromes. A number of papers were published on these studies. From the human studies he went on to study the embryonic development of 10-day to term mouse specimens with trisomy 19 or trisomy 16—a model that has been used for the Down syndrome. The studies included morphologic development of the placentas and patterns of expression of major histocompatability antigens in affected and normal littermate conceptuses, as well as the classic embryonic development of the conceptuses.

Rui Diogo is a multi-awarded (American Assoc. Anatomists, Anatomical Soc. Great Britain & Ireland) Assistant Professor at the Howard Univ. College of Medicine and a Resource Faculty at the Center for the Advanced Study of Hominid Paleobiology of George Washington Univ. (US). He is the author/co-author of numerous publications, and co-editor of the books *Catfishes* and *Gonorynchiformes and ostariophysan interrelationships*. He is the sole author or first author of several monographs, including five atlases of baby gorillas and adult gorillas, chimpanzees, hylobatids and orangutans and the books *Morphological evolution, aptations, homoplasies, constraints and evolutionary trends, The origin of higher clades - osteology, myology, phylogeny and evolution of bony fishes and the rise of tetrapods, Muscles of vertebrates*, and *Comparative anatomy and phylogeny of primate muscles and human evolution*.

Color Plate Section

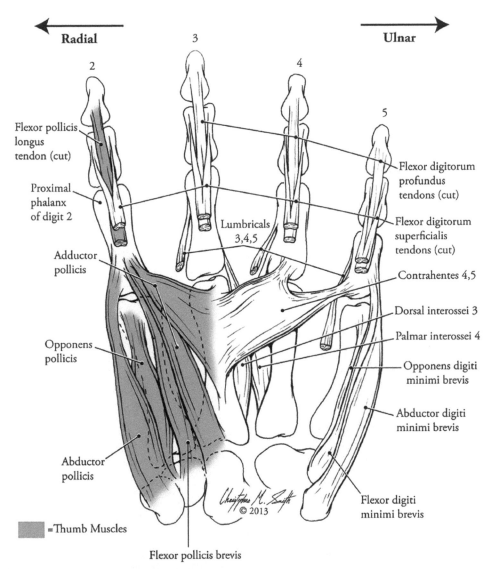

Radial ←   3   → **Ulnar**

2

4

5

Flexor pollicis longus tendon (cut)

Proximal phalanx of digit 2

Adductor pollicis

Lumbricals 3,4,5

Flexor digitorum profundus tendons (cut)

Flexor digitorum superficialis tendons (cut)

Contrahentes 4,5

Dorsal interossei 3

Palmar interossei 4

Opponens pollicis

Opponens digiti minimi brevis

Abductor digiti minimi brevis

Abductor pollicis

© 2013

Flexor digiti minimi brevis

= Thumb Muscles

Flexor pollicis brevis

Figure 1.4. Scheme illustrating the thumb muscles of the other hand (the one with six digits) of the same human newborn with Trisomy 18. Our hypothesis is supported because, despite the presence of two thumbs (which are the only digits illustrated in this figure), the muscles normally associated with the thumb are not duplicated. Instead, the muscles that normally insert respectively onto the radial and ulnar sides of the thumb insert onto the radial and ulnar sides of the most radial and most ulnar thumbs, respectively, as predicted. Interestingly, the tendon of the flexor digitorum profundus, which usually goes to the central (so, not ulnar and not radial) portion of the thumb, bifurcates to go to both the most ulnar and most radial thumbs.

Normal

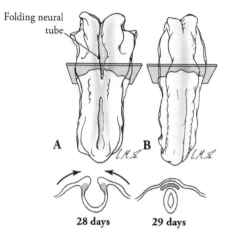

Folding neural tube

Neural crest cells originate from the "crest" of the folding neural tube and reside between the neural tube and ectoderm after fusion

28 days   29 days

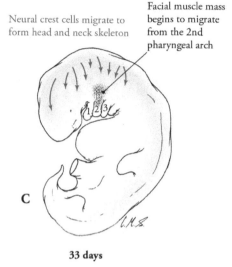

Neural crest cells migrate to form head and neck skeleton

Facial muscle mass begins to migrate from the 2nd pharyngeal arch

33 days

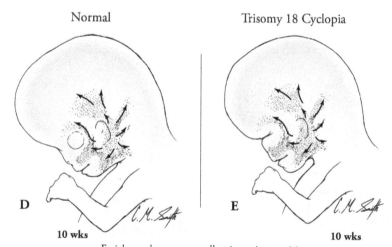

Normal   Trisomy 18 Cyclopia

10 wks   10 wks

Facial muscle precursor cells migrate into position across the face attaching onto neural crest cell derived structures and then inserting into skin and cartilage

Figure 1.5. Development of facial musculature. Normal development with neural crest cells depicted in purple, at 28 days **(A)**, 29 days **(B)**, 33 days **(C)**. Comparison of normal **(D)** and Trisomy 18 cyclopic **(E)** facial muscle migration at 10 weeks. Topological position may play a role in facial muscle attachments, as seen from this study (see Chapter 4). Our hypothesis is that in both normal and Trisomy 18 cases, the facial muscles migrate essentially in the same directions, regardless of underlying skeletal structure, then mainly attach to their "nearest neighbor" based on their topological position in space.

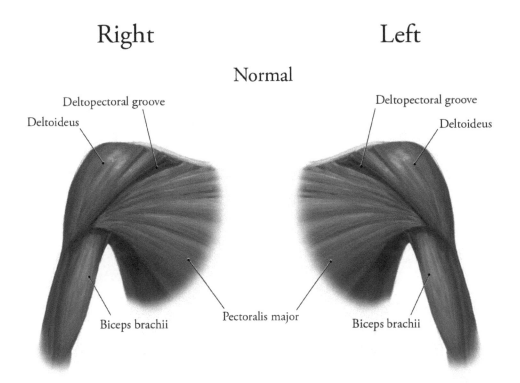

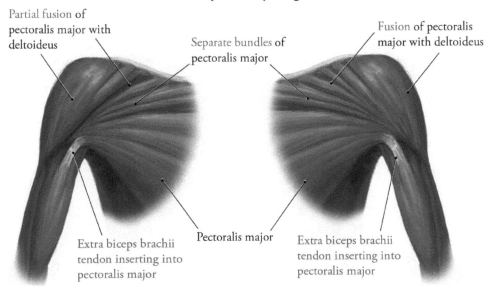

Figure 2.1. Comparison of shoulder musculature in normal and Trisomy 18 cyclopia. Anomalies shown in red.

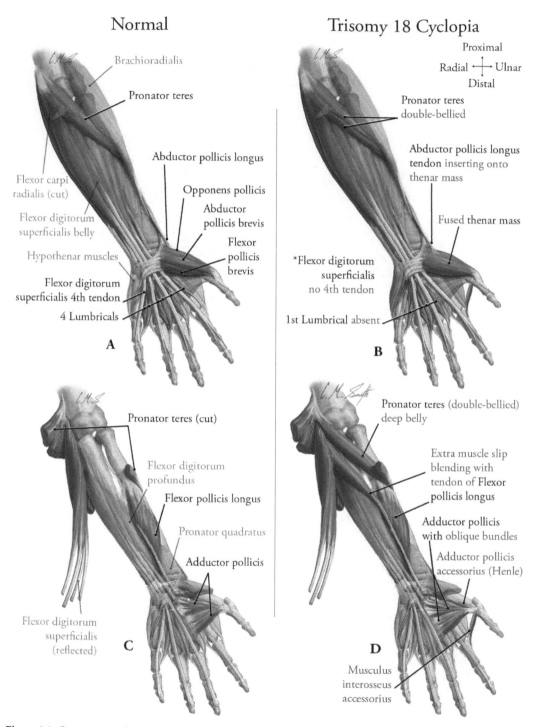

Figure 2.2. Comparison of normal and Trisomy 18 cyclopia left flexors. Superficial muscles (**A, B**), deep muscles (**C, D**). Black labels on A and C indicate normal presentation of structures found to be anomalous in Trisomy 18 cyclopia (B, D). Anomalous features (red labels), cartilage (purple), normal structures included for orientation (grey labels).

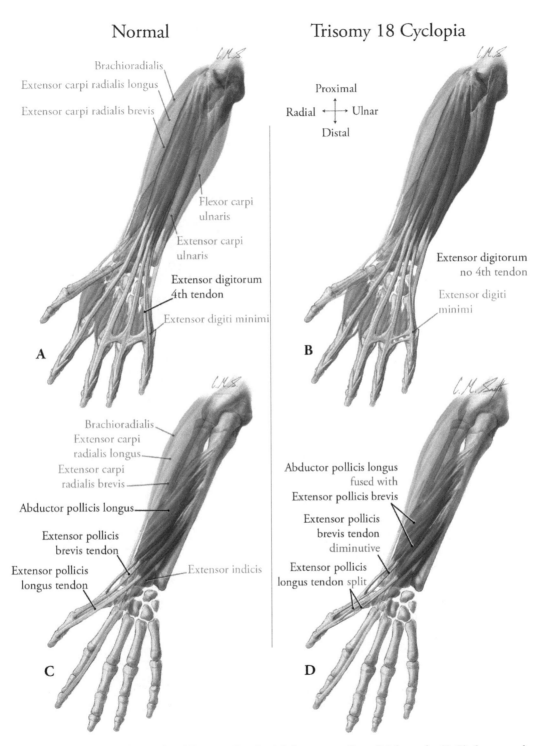

Figure 2.3. Comparison of normal and Trisomy 18 cyclopia left extensors. Superficial muscles **(A, B)**, deep muscles **(C, D)**. Black labels on A and C indicate normal presentation of structures found to be anomalous in Trisomy 18 cyclopia (B, D). Anomalous features (red labels), cartilage (purple), normal structures included for orientation (grey labels).

Trisomy 18 Cyclopia Left Forearm and Hand

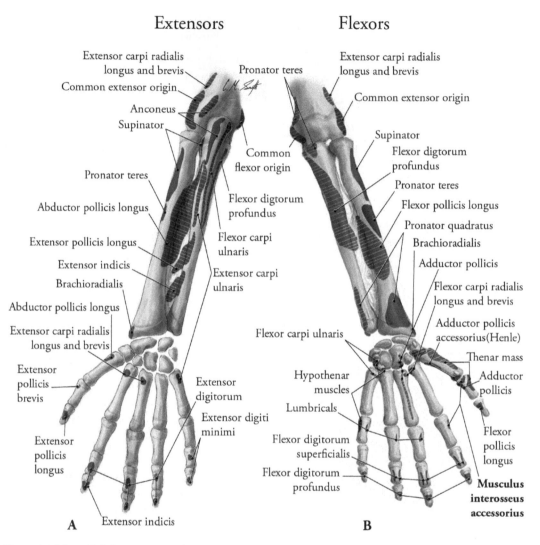

Figure 2.4. Map of left forearm musculature origins and insertions in Trisomy 18 cyclopia. Origins are shown in red (striped). Insertions are shown in blue. Non-ossified cartilage is shown in purple.

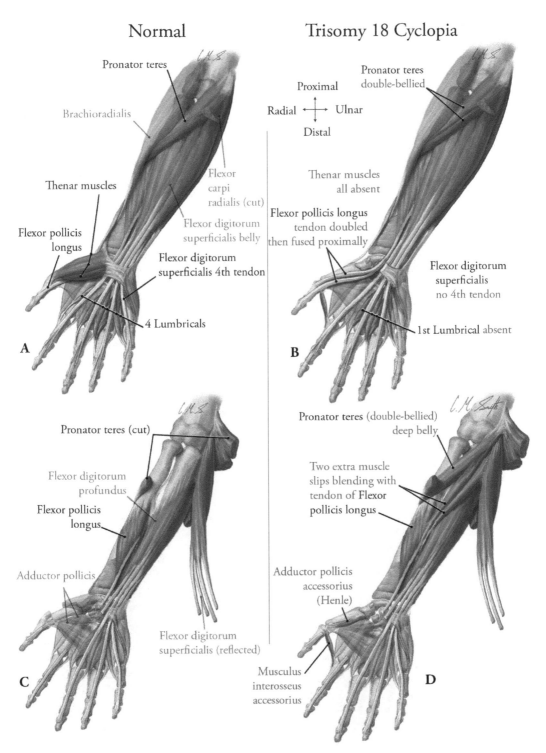

Figure 2.5. Comparison of normal and Trisomy 18 cyclopia right flexors. Superficial muscles **(A, B)**, deep muscles **(C, D)**. Black labels on A and C indicate normal presentation of structures found to be anomalous in Trisomy 18 cyclopia (B, D). Anomalous features (red labels), cartilage (purple), normal structures included for orientation (grey labels).

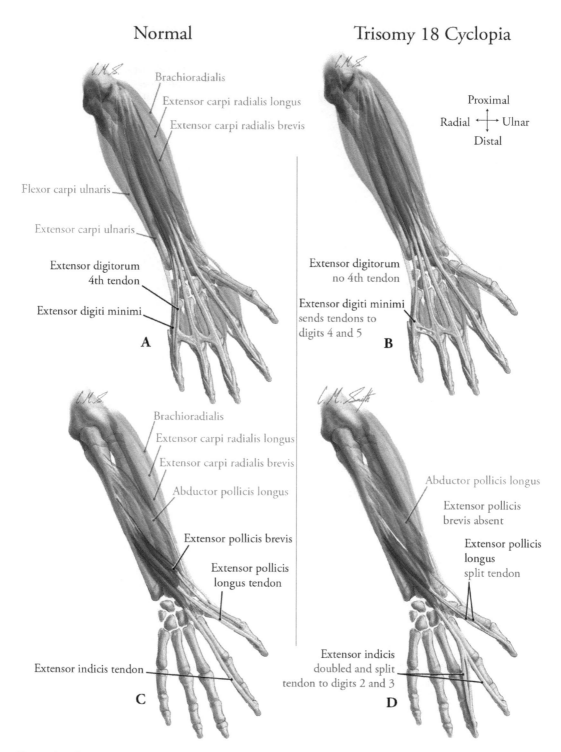

Figure 2.6. Comparison of normal and Trisomy 18 cyclopia right extensors. Superficial muscles **(A,B)**, deep muscles **(C, D)**. Black labels on A and C indicate normal presentation of structures found to be anomalous in Trisomy 18 cyclopia (B, D). Anomalous features (red labels), cartilage (purple), normal structures included for orientation (grey labels).

Trisomy 18 Cyclopia Right Forearm and Hand

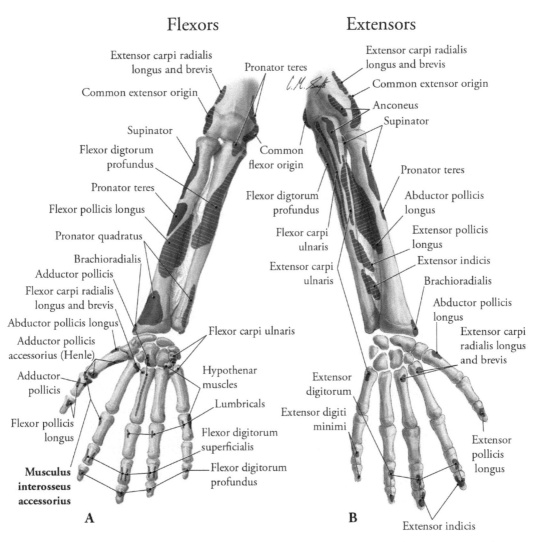

Figure 2.7. Map of right forearm musculature origins and insertions in Trisomy 18 cyclopia. Origins are shown in red (striped). Insertions are shown in blue. Non-ossified cartilage is shown in purple.

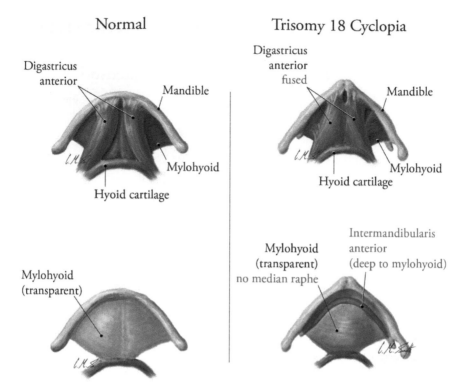

Figure 2.8. Comparison of inferior mandibular view in normal and Trisomy 18 cyclopia. Anomalies labeled in red. Cartilage shown in purple.

Normal

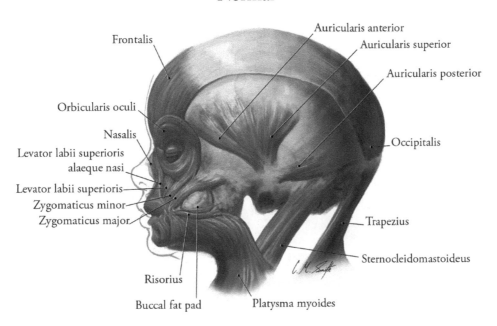

Frontalis

Auricularis anterior

Auricularis superior

Auricularis posterior

Orbicularis oculi

Nasalis

Levator labii superioris alaeque nasi

Levator labii superioris

Zygomaticus minor

Zygomaticus major

Occipitalis

Trapezius

Sternocleidomastoideus

Risorius

Buccal fat pad

Platysma myoides

Trisomy 18 Cyclopia

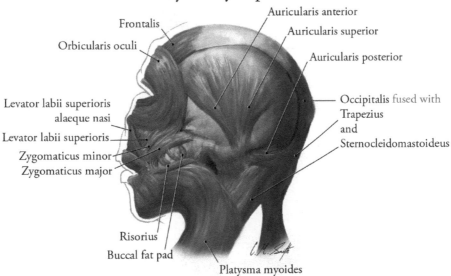

Frontalis

Auricularis anterior

Auricularis superior

Orbicularis oculi

Auricularis posterior

Levator labii superioris alaeque nasi

Levator labii superioris

Zygomaticus minor

Zygomaticus major

Occipitalis fused with Trapezius and Sternocleidomastoideus

Risorius

Buccal fat pad

Platysma myoides

Figure 2.9. Comparison of superficial lateral head musculature in normal and Trisomy 18 cyclopia. Presence of a platysma cervicale could not determined due to previous dissection.

Normal

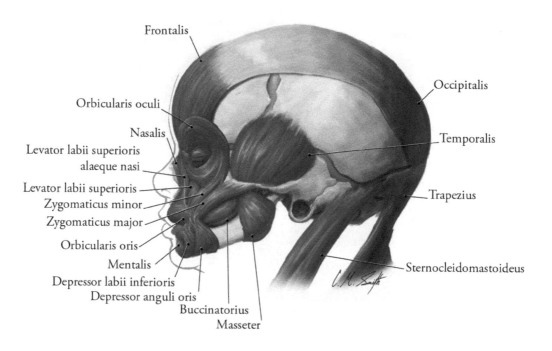

Trisomy 18 Cyclopia

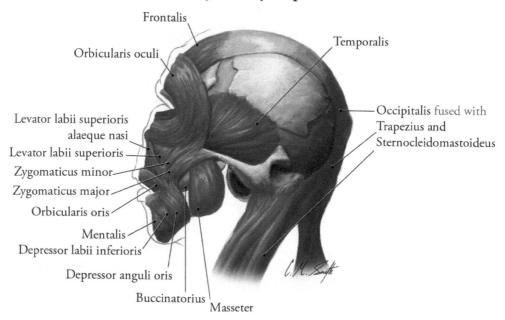

Figure 2.10. Comparison of lateral head musculature in normal and Trisomy 18 cyclopia. Platysma myoides, risorius, and buccal fat pad removed.

Normal

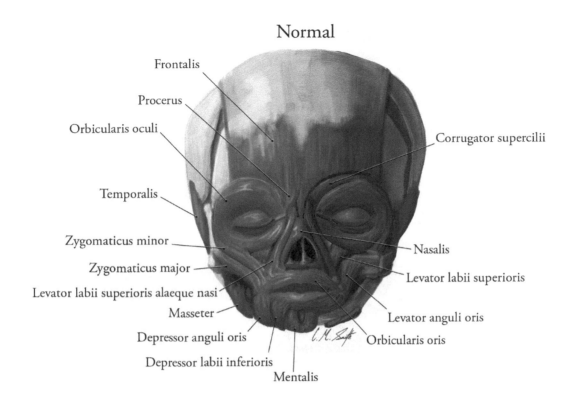

Frontalis

Procerus

Orbicularis oculi

Corrugator supercilii

Temporalis

Zygomaticus minor

Zygomaticus major

Levator labii superioris alaeque nasi

Masseter

Depressor anguli oris

Depressor labii inferioris

Mentalis

Nasalis

Levator labii superioris

Levator anguli oris

Orbicularis oris

Trisomy 18 Cyclopia

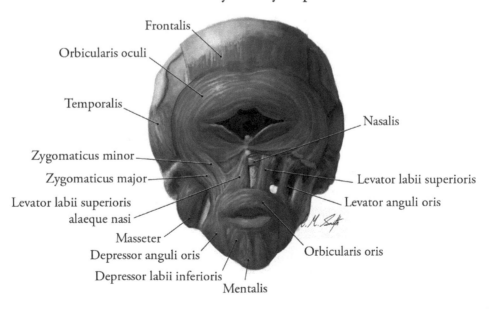

Frontalis

Orbicularis oculi

Temporalis

Zygomaticus minor

Zygomaticus major

Levator labii superioris alaeque nasi

Masseter

Depressor anguli oris

Depressor labii inferioris

Mentalis

Nasalis

Levator labii superioris

Levator anguli oris

Orbicularis oris

Figure 2.11. Comparison of anterior head musculature in normal and Trisomy 18 cyclopia. Platysma myoides, risorius, and buccal fat pad removed. Left side shows deep dissection.

Trisomy 18 Cyclopia Lateral Skull

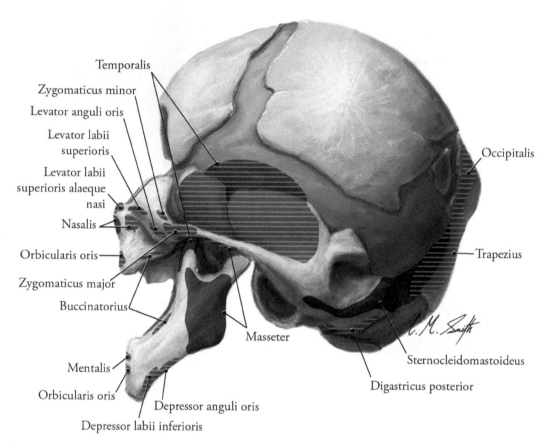

Figure 2.12. Map of lateral skull origins and insertions in Trisomy 18 cyclopia. Origins are shown in red (striped). Insertions are shown in blue. Fontanelles are shown in purple.

Trisomy 18 Cyclopia Anterior Skull

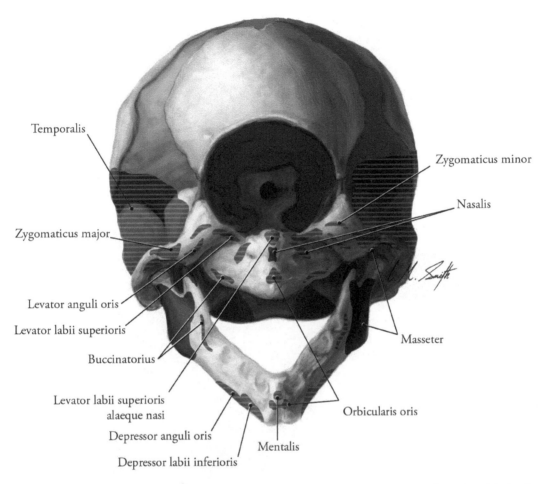

Figure 2.13. Map of anterior skull origins and insertions in Trisomy 18 cyclopia. Origins are shown in red (striped). Insertions are shown in blue. Fontanelles are shown in purple.

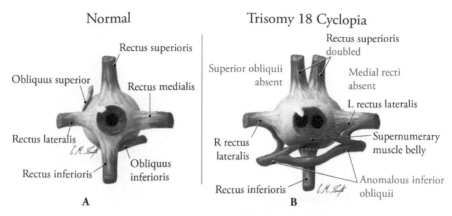

Normal   Trisomy 18 Cyclopia

Figure 2.14. Comparison of normal and Trisomy 18 cyclopia eyes. Orbit musculature with cut bellies reflected anteriorly. Normal right eye **(A)**, contrasted with fused irises in Trisomy 18 cyclopia **(B)** Anomalies labeled in red.

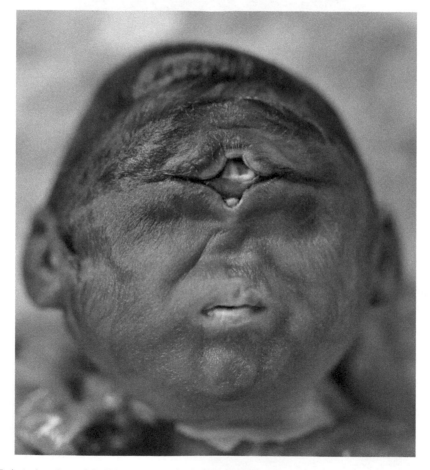

Figure 2.15. Anterior view of the Trisomy 18 cyclopia face. Note the interesting upper eyelid structure of two lids and medial portion.

Normal

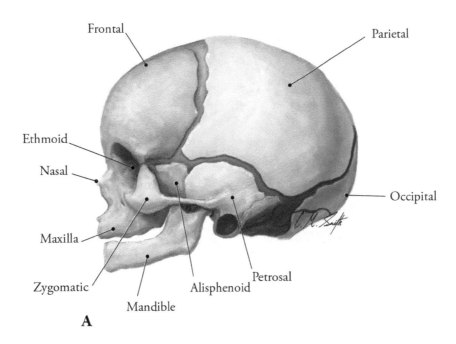

A

Trisomy 18 Cyclopia

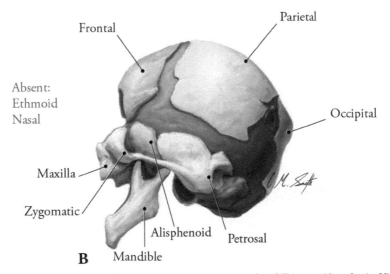

B

Figure 2.16. Lateral comparison of the bones of the skull in normal and Trisomy 18 cyclopia. Visible absent bones are labeled in red.

Normal

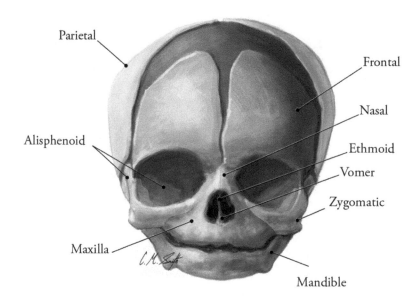

Trisomy 18 Cyclopia

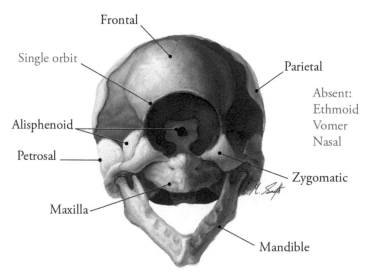

Figure 2.17. Anterior comparison of the bones of the skull in normal and Trisomy 18 cyclopia. Visible absent bones are labeled in red.

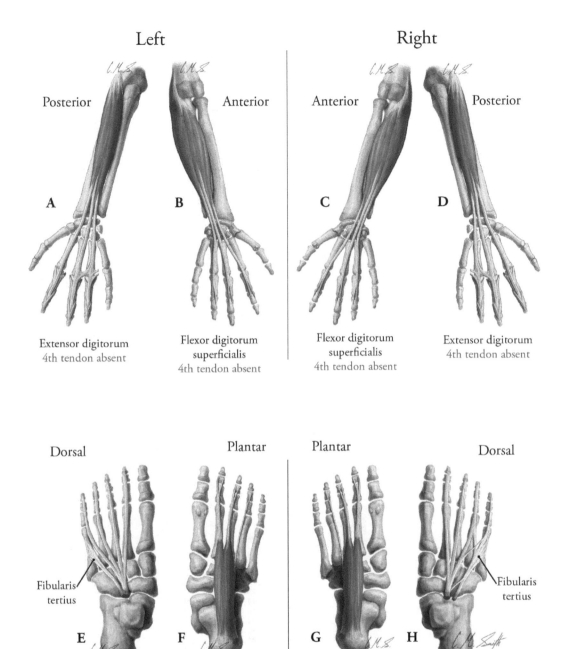

Figure 4.1. Trisomy 18 cyclopia upper and lower limb comparisons. Red indicates the tendon absent in all four limbs: 4th tendon to the 5th digit (**A-D, F, G**). Normal anatomy was observed on the dorsal surface of the left foot (**E**). The 4th tendon inserts onto separate fibularis tertius on the dorsal surface of the right foot (**H**).

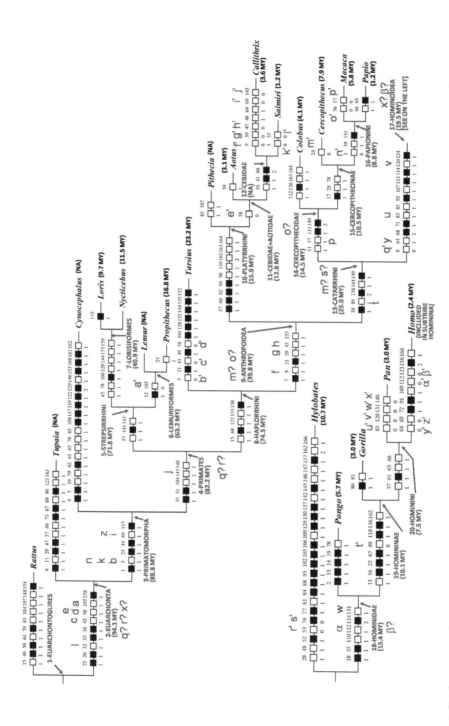

Figure 6.1. Single most parsimonious tree (L 301, CI 58, RI 73) obtained from the analysis of of 166 characters of the head, neck, pectoral and forelimb musculature (Diogo and Wood, 2011; 2012a). The unambiguous transitions that occurred in each branch are shown in white (homoplasic transitions) and black (non-homoplasic transitions) squares (numbers above and below the squares indicate the character and character state, respectively). Together with the name of each euarchontan clade is shown the respective estimate molecular divergence time, excepting for the genus *Homo* for which it is shown a time of origin exclusively based on the fossil record (see text for more details). A detailed description of the 28 unambiguous reversions to a plesiomorphic state is given in Diogo and Wood, 2012b; in the text of the present Chapter we only refer to some of these reversions (N.B., letters/symbols without a prime indicate the nodes where the respective original transitions from the plesiomorphic state to the derived state took place).

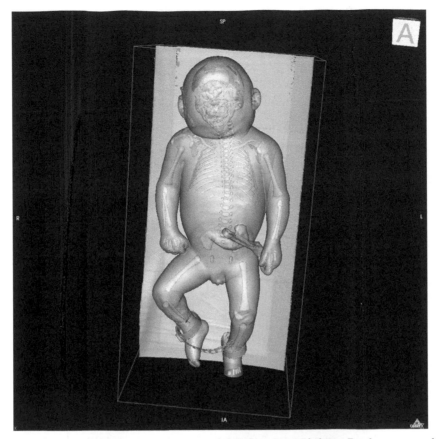

Figure B.1. Anterior view surface render. Transparent skin is shown over skeleton. Render was completed using Osirix DICOM viewer.

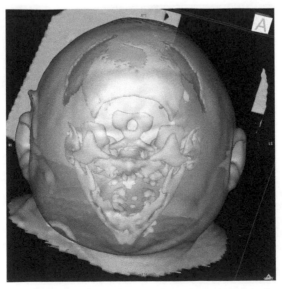

Figure B.3. Anterior view surface render of the cyclopia head. Transparant skin is shown over skull. Render was completed using Osirix DICOM viewer.

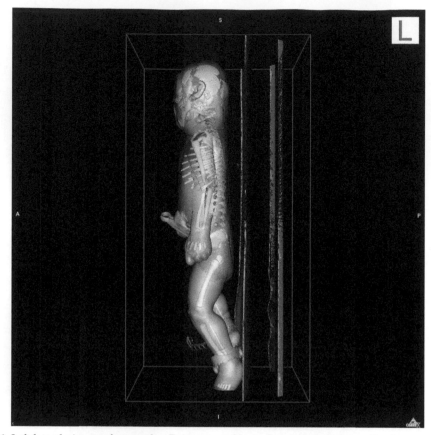

Figure B.6. Left lateral view surface render. Transparent skin is shown over skeleton. Render was completed using Osirix DICOM viewer.

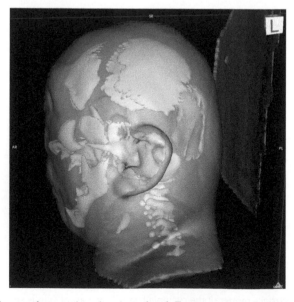

Figure B.8. Left lateral view surface render of cyclopia head. Transparent skin is shown over skull. Render was completed using Osirix DICOM viewer.

Milton Keynes UK
Ingram Content Group UK Ltd.
UKHW050451071024
449327UK00015B/336